PLANT TROPISMS AND OTHER GROWTH MOVEMENTS

PLANT TROPISMS AND OTHER GROWTH MOVEMENTS

J. W. HART

Formerly of the
Department of Plant and Soil Science,
University of Aberdeen

CHAPMAN & HALL

London · Glasgow · New York · Tokyo · Melbourne · Madras

Published by Chapman & Hall, 2–6 Boundary Row, London SE1 8HN

Chapman & Hall, 2–6 Boundary Row, London SE1 8HN, UK

Blackie Academic & Professional, Wester Cleddens Road, Bishopbriggs, Glasgow G64 2NZ, UK

Chapman & Hall, 29 West 35th Street, New York NY10001, USA

Chapman & Hall Japan, Thomson Publishing Japan, Hirakawacho Nemoto Building, 6F, 1-7-11 Hirakawa-cho, Chiyoda-ku, Tokyo 102, Japan

Chapman & Hall Australia, Thomas Nelson Australia, 102 Dodds Street, South Melbourne, Victoria 3205, Australia

Chapman & Hall India, R. Seshadri, 32 Second Main Road, CIT East, Madras 600 035, India

First edition 1990
Reprinted 1992

© 1990 J.W. Hart

Typeset in 10/12pt Palatino by Computape (Pickering) Ltd, North Yorks
Printed in Great Britain at the University Press, Cambridge

ISBN 0 412 53080 5

A catalogue record for this book is available from the British Library

Library of Congress Cataloging-in-Publication data available

As always,
to
Leona

If we look, for instance, at a great acacia tree, we may feel assured that every one of the innumerable growing shoots is constantly describing small ellipses; as is each petiole, sub-petiole, and leaflet . . . The flower peduncles are likewise continually circumnutating . . . All this astonishing amount of movement has been going on year after year since the time when, as a seedling, the tree first emerged from the ground.

Charles Darwin (1880)
The power of movement in plants

No plant is entirely without the power of movement . . . The fact that in large plants the power of growth and movement are not strikingly evident has caused plants to be popularly regarded as 'still life'. If mankind . . . were accustomed to view nature under a magnification of 100 to 1,000 times, or to perceive the activities of weeks or months performed in a minute, as is possible by the aid of a kinematograph, this erroneous idea would be entirely dispelled.

Wilhelm Pfeffer (1906)
The physiology of plants

Preface

Life involves movement. Plants are no exception to this generality, although their mechanisms of movement are markedly different from those of animals. One way in which even sedentary higher plants move is through changes in the rates or patterns of growth within their organs, often directly in response to stimulation by some environmental factor. Tropisms are one such form of plant movement, and represent a major means by which a plant senses, responds, and adjusts to its surroundings. This book attempts to give an account of this area, and to demonstrate that the beauty, and sensitivity, of plants extends beyond any static 'still-life' picture.

Chapters 1 and 2 are introductory, dealing respectively with plant movements in general and tropisms in particular. The next three chapters deal with responses to the major types of tropic stimuli, i.e. gravity, light, and mechanical factors. And the final chapter considers other types of stimuli, such as chemicals, water, and injury, which are of most significance only to particular plants, particular stages of development, or particular situations. Throughout the text there are also special topic 'boxes' consisting of squared off sections of text. These are simply devices for presenting explanatory background information, or for considering related aspects that I myself find particularly intriguing.

In the interests of space and readability no attempt is made to produce an exhaustive literature review, particularly with regard to the older literature. In many cases, of course, some specificity or priority is required and original sources are given, but in general, access to the older literature can be obtained through the reviews that are cited in the text in relation to particular areas.

I would like to thank Norman Little for his expert assistance with many of the illustrations and photographs, and my typist, Mrs Betty Smith, for expertly disentangling and dealing with my manuscript.

J. W. Hart
Aberdeen, 1989

Acknowledgements

We are grateful to the following individuals and organizations who have kindly given permission for the reproduction of copyright material (figure numbers in parentheses):

A. Johnsson and Gustav Fischer Verlag (1.3b); Blackwell Scientific Publications (1.4, 3.9, 4.12); V. E. A. Russo and Plenum Publishing (2.1, 4.4); W. K. Silk, reproduced by permission from *Annual Review of Plant Physiology* 35, © 1984 Annual Reviews Inc.; Springer-Verlag (3.1, 4.3, 5.4(a,b), 6.1); DLG-Verlag (3.3); A. Sievers and Springer-Verlag (3.6, 3.7); A. Sievers, J. M. Selker and the Editor, *American Journal of Botany* (3.8); M. B. Wilkins and Longman (3.11); H. Mohr and Springer-Verlag (4.1); D. S. Dennison and Springer-Verlag (4.2a, 4.8a); M. Black and Springer-Verlag (4.5); T. I. Baskin, reproduced by permission from Baskin and Iino, *Petrochemistry & Photobiology* 46, © 1986 Pergamon Press PLC (4.6a); American Society of Plant Physiologists (4.6b); G. Blaauw-Jansen and the Editor, *Acta Botanica Neerlandica* (4.8b); W. R. Briggs, © 1957 American Association for the Advancement of Science (4.10); T. C. Rich and Blackwell Scientific Publishing (4.11); H. van den Erde and Academic Press (6.2, 6.3); the Editor, *American Journal of Botany* (6.4).

Contents

List of special topic boxes

xv

List of tables

CHAPTER ONE

Tropisms and other forms of plant movement

Tropisms constitute one particular form of directional movement shown by plants. Before the detail of later chapters, it may be useful to consider them within this general context of plant movement in order to get some idea of their overall significance in the life of the plant.

This approach also provides the opportunity to introduce turgor responses, another major form of plant movements that are sometimes closely associated with tropisms.

1.1 PLANTS AND MOVEMENT

If asked to distinguish between plants and animals, perhaps most people would sooner or later offer the observation that 'animals move and plants do not'. But in fact, movement pervades all aspects of plant behaviour, though not always in the restricted sense of locomotion of the whole organism.

In the plant kingdom, controlled voluntary locomotion is seen only in the streaming (slime moulds), gliding (diatoms), and free-swimming movements (flagellated algae) of lower plants and certain kinds of reproductive cells. The types of protein-based mechanisms that are responsible for these movements are restricted in higher plants to driving intra-cellular movements, such as occur in cell division, cytoplasmic streaming, and so forth.

In higher plants, movement of the whole organism does occur in one sense, through the dispersal of reproductive units such as seeds, spores, and vegetative propagules. However, in all the wide variety of mechanisms used in this area, whether based on some form of explosive dehiscence or on passive transport by wind, water, or animals, the energy for the actual movement itself does not come directly from the plant's own metabolic activity, but is supplied by some other agency or

1

organism. Therefore, although reproductive dispersal can result in the movement of plants over vast distances, it represents an involuntary form of movement that is neither under the direct control of the plant nor necessarily related to immediate environmental problems or pressures.

All higher plants move, however, in the sense that they are continually carrying out controlled changes in the orientations and juxtapositions of their various parts. Such movements, directly analogous to changes of posture in animals, are brought about by two distinct types of mechanisms in plants. *Turgor movements* are due to reversible changes in the sizes of special cells which bring about the movement of a lever arm, such as a leaf blade or a reproductive structure. *Growth movements* result from particular patterns of differential growth within or between organs, and, although necessarily slow, are no less dramatic or crucial to the life of the plant.

Each of these forms of movement, by change in turgor or by change in growth, can be controlled endogenously or exogenously. Endogenously regulated movements are initiated or controlled by some factor within the plant, although factors in the external environment may have an indirect effect on the movement. In most cases, the exact nature of the internal regulatory factor is not clear. Endogenously regulated turgor movements generally occur in some kind of periodic repeating pattern or rhythm. When the rhythm is approximately the length of a day, the movements are considered to be controlled by, or to be a manifestation of, the circadian clock (Koukkari & Warde 1985). Movements showing much shorter rhythms are known, but in these the regulatory mechanism is completely obscure. Endogenously regulated growth movements are generally part of the normal pattern of development of an organ and, as such, are considered to derive from some preprogrammed physiological asymmetry within the tissues, perhaps, for example, involving some sort of differential behaviour of a growth-regulating chemical or some sort of differential sensitivity of the tissues towards a growth regulator.

In exogenously regulated movements some environmental factor *directly* initiates or controls the change in turgor or growth rate that is responsible for the movement. Turgor movements are generally initiated either by light or by physical contact, depending on the type of movement. Directed growth movements are most generally controlled by light or by gravity. However, it is becoming increasingly apparent that other forms of mechanical stimulation, such as contact, flexure, vibration, and pressure, also play a major role in regulating growth movements. And, particularly in lower plants, specific chemical signals often stimulate growth movements. Temperature, of course, also exerts major effects on plant growth. Thus, the range and types of environmental factors that initiate or control plant movements are directly comparable to those

2

Table 1.1 Summary of the major forms of movement responsible for changes in the orientation of plant organs.

Type of stimulus	Type of mechanism	
	Turgor change	Growth
endogenous	nyctinasty* (leaves) ultradian rhythms (leaves) flower movement* (sunflower)	nutations (all organs) epinasty (petioles, shoots) autotropism (see Ch. 2)
temperature		thermonasty (petals)
light	photonasty (leaves) heliotropism (leaves)	photonasty† (flowers) phototropism† (aerial organs)
gravity		gravitropism (most organs)
mechanical	seismonasty (*Mimosa* leaves) thigmonasty (floral parts) (insect traps)	thigmonasty (most tendrils) thigmotropism (some tendrils)
chemical		epinasty (petioles, shoots) chemotropism (fungal hyphae, pollen tube?)
injury		traumatropism (all organs?)

* The rhythms of these movements are entrained by light.
† In a nastic movement the direction of response is determined by features of the tissue; in a tropic movement, the direction of response is determined by the stimulus.

provoking sensory responses in animals and, indeed, in many instances the sensitivity of a plant or plant part towards any one of these environmental factors is of the same order of magnitude as that of many animals (Shropshire 1979). (It is only in regard to their responsiveness towards external electrical stimulation that the sensory ranges of plants and animals seem to show any appreciable difference, although even in this respect it is now clear that, internally, electrical signals are important within the plant (see Ch. 2), and, externally, in some responses (Miller *et al.* 1986) electrical signals may act as important sources of directional information to certain types of plant.)

All of these regulatory environmental factors initiate and control plant movements by acting through catenary events, which involve processes of stimulus reception, physiological mediation of the signal, and regulation of the eventual biological response. These aspects are discussed in Chapter 2.

In the nomenclature of plant movements, the two terms 'nastic' and 'tropic' (or 'tropistic') are commonly used. It should be noted that these terms do not relate to the particular mechanism of movement (turgor change or growth response), nor specify the means of regulation (endogenous or exogenous). They simply refer to the directionality of the movement. A nastic response is one in which the direction of the movement is determined by features of the tissue, rather than by the stimulus; thus, nastic movements can be initiated by endogenous or exogenous factors; and they can also result from turgor changes or from growth. A tropic response, on the other hand, is one in which the direction of movement is strongly related to the direction of some environmental factor; thus tropic responses are always initiated by exogenous stimuli; they usually result from changes in growth, though some (e.g. tendril movements) may also involve turgor responses and at least one (i.e. heliotropism) seems to result wholly from turgor changes.

The various types of turgor and growth movements are summarized in Table 1.1, and a general account of each of them is given in the rest of this chapter.

Box 1.1   Movement and life's problems.

All organisms, including plants, animals, and humans, must overcome problems in three general areas in order to be biologically successful. These are:

(a) *Self-protection*, which involves protection against climatic conditions and against other organisms.

(b) *Food collection*, which in its widest sense includes processes involved in reaching (or creating) an environment that provides adequate levels of resources, as well as processes involved in the actual uptake of food into the organism.

(c) *Reproduction*, which, to be wholly successful, involves not only fertilization and the production of offspring, but also the dispersal and establishment of the offspring in suitable environments.

Of course, most animals utilize some form of voluntary locomotion or movement in all of these areas, but sedentary higher plants must depend largely upon other means. These are often based upon the elaboration of specialized metabolic pathways, e.g. chemical methods of protection (against other organisms by means of toxins, or against high light intensities by means of pigments), and chemical enhancement of reproductive efficiency (by the production of visual or olfactory attractants). And they are also often based upon specialized modes of development, e.g. protection against the rigours of climate by means of dormancy, and regulation of reproductive events by means of photoperiodism.

However, as should become apparent from the discussion throughout this and subsequent chapters, plants do use movement in various forms and to different extents in each of these biological 'problem areas'. For example, the relative rapidity of turgor movements means that they are often involved in aspects of protection and reproduction. On the other hand, the slower growth movements; and in particular the directionally controlled tropisms, are more generally involved in 'moving' plant parts towards the most favourable environment.

1.2   PLANT TURGOR MOVEMENTS

1.2.1   General nature of the mechanism

Turgor movements are brought about by osmotically driven changes in the volume of special cells, or of cells in special regions (Hill & Findlay 1981, Findlay 1984). That is, the osmotic potential of the cells is increased, usually by the uptake of potassium and balancing anions such as Cl^-; water is therefore taken in and the cells increase in size. When potassium is lost from the cells, water is also lost and the cells decrease in size. The uptake of potassium is generally considered to be an active, energy-requiring process, perhaps coupled to proton extrusion through a membrane ATP-ase, but its efflux is thought to be passive (Satter 1979, Galston 1983). However, in some extremely rapid turgor movements, such as occur in certain pollination mechanisms, the phase of cell shrinkage is so fast that some contractile mechanism is also thought to be involved in the active expulsion of water (Findlay 1984).

The volume changes of the guard cells during opening and closing of stomata represent a structurally simple form of turgor movement, but the same type of osmotic engine is also used to drive movements that are more elaborately structured and more obvious. For example, in many species the base of the leaf petiole consists of a bulbous region, the *pulvinus*, which contains special, structurally distinct 'motor cells', with very thin walls. Turgor changes in the motor cells bring about changes in the angle of the leaf petiole, i.e. the pulvinus functions as a kind of active hinge. The anisotropic changes in volume of the motor cells in *Phaseolus coccineus* seem to be due to the hoop-like arrangement of cellulose microfibrils in their longitudinal walls (Mayer *et al.* 1985).

In many types of turgor movements, however, the arrangement and appearance of the motor cells are not so anatomically distinct as in the leaf pulvinus.

1.2.2   Types of turgor movements

Turgor movements can be somewhat arbitrarily categorized according to whether the major factor that is directly responsible for their initiation is of endogenous origin or is some exogenous environmental stimulus.

Endogenously regulated turgor movements
The leaves of many plants, including all legumes, show daily changes in orientation known as 'sleep movements' or *nyctinastic* movements (literally, 'night-folding'). In these movements (Fig. 1.1a) the leaves are opened out during the day and folded in along the stem axis at night, upwards or downwards depending on the species. (The adaptive sig-

(a)

(b)

Key
F flexor cells (swell during closure)
E extensor cells (swell during opening)
⟶ directions of K⁺ movement

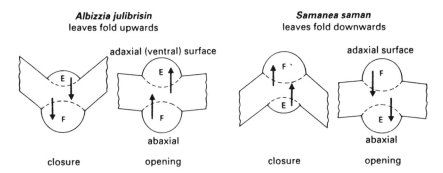

Albizzia julibrisin
leaves fold upwards

adaxial (ventral) surface

abaxial

closure   opening

Samanea saman
leaves fold downwards

adaxial surface

abaxial

closure   opening

Figure 1.1 Nyctinastic sleep movements in leaves. (a) Leaf orientation in *Desmodium gyrans* (the 'telegraph plant') during the day (left) and at night (from Darwin 1880). (b) Summary of events during sleep movements in two species of legume, viewed through idealized sections of rachis and leaflets (from Hart 1988).

nificance of these movements is not too clear, but they may minimize radiative heat loss from the plant at night.) In some cases, these leaf movements may be due to changes in growth rate (Yin 1941), but they more generally result from rhythmic turgor changes in the pulvinar motor cells (Satter 1979, Galston 1983). In many species these motor cells consist of two functionally distinct types (Fig. 1.1b): during leaf opening, those designated as 'extensor' cells take up potassium and swell, and the 'flexor' cells lose potassium and decrease in size; during leaf folding, the directions of potassium movement are reversed and the extensors shrink while the flexors swell.

Nyctinastic leaf movements show a strong circadian rhythm, i.e. they continue in a cycle of approximately 24 hours even under constant environmental conditions and continuous darkness. The actual basis of circadian timing, the so-called biological clock, is not yet clear, although several models have proposed a central role for the cell membrane in the timing mechanism (reviewed in Engelman & Schrempf 1980). For example, one such model (Njus *et al.* 1974) suggests that the timing system involves the active pumping of potassium into some cell compartment until at saturation the pump is switched off by feedback inhibition; the circadian timing period then results from the time taken for potassium to passively leak out of the compartment and the pump to become switched on again. Such events may be involved in the turgor changes of nyctinasty.

However, a certain class of chemicals has also been shown to specifically regulate leaf turgor movements in a wide range of species (Schildknecht 1984). These chemicals, derivatives of organic acids such as gallic acid and catechuic acid, have received the general names 'turgorins' or 'leaf movement factors' (LMFs) and are thought to bring about water loss from the motor cells by binding to the cell membrane. The turgorin LMF1, from *Robinia pseudoacacia*, has also been shown to induce stomatal closure in *Commelina communis* (Paterson *et al.* 1987). It has been suggested that rhythmic cycling of turgorins between active and inactive forms may be involved in circadian timing (Schildknecht 1984).

Circadian leaf movements can also be influenced by factors in the external environment, particularly light. The actions of two separate photoreceptors seem to be involved (Galston 1983). Blue light, given at the appropriate stage of the circadian rhythm, generally stimulates leaf opening, and in some species this effect completely overrides the rhythm, to result in photonastic and heliotropic leaf movements (discussed in the next section). The effects of red light, acting through phytochrome are less straightforward and seem to involve effects on the entrainment of the circadian rhythm as well as effects on the actual rates and extents of leaf opening (Galston 1983). In a few species, mechanical stimulation can also override the circadian rhythm and induce leaf closure (discussed on pp. 9–11).

Other types of endogenously regulated leaf movements are also present in many species. These often occur as short-term (ultradian) rhythms, superimposed on the diurnal cycle. The lengths of the rhythms vary according to species. They can occur in cycles of several hours, e.g. in peanut (*Arachis hypogea*) the leaves move up and down over a 5 hour cycle, in clover (*Trifolium subterrana*) over a $3\frac{1}{2}$ hour cycle, and in wood sorrel (*Oxalis acetosella*) over a $1\frac{1}{2}$ hour cycle (Darwin 1880). Again, some leaf movements occur in cycles of minutes, e.g. in cotton (*Gossypium hirsutum*) 36 minutes, and in bean (*Phaseolus vulgaris*) 53 minutes (Kouk-kari & Warde 1985). And in the appropriately named 'telegraph plant' (*Desmodium gyrans*, Fig. 1.1a), the subsidiary leaflets show vigorous circling movements over periods of $1\frac{1}{2}$–3 minutes depending on the temperature (Coombe & Bell 1965). Neither the underlying mechanisms, nor the significance, of this astonishingly wide variety of leaf movements are at all clear.

Many flowers also show daily rhythms of opening and closing. However, since the mechanisms responsible for these movements usually involve considerable amounts of growth and a greater degree of environmental control, they are discussed in Section 1.3.2.

One case of flower movement, however, that does seem to be largely under endogenous control is the continuous reorientation shown throughout the day by the flower heads of sunflower (*Helianthus annuus*). Rather than being an example of heliotropism, or sun-tracking, this movement seems to be another manifestation of the circadian rhythm (Leshen 1977). The extents to which changes in cell turgor and changes in growth rate contribute to the movement are not entirely clear.

Exogenously induced turgor movements

Many types of turgor movements, even some of the strongly rhythmic movements, can be triggered by the actions of environmental factors, particularly light and mechanical stimuli.

Some shade species, e.g. wood sorrel (*Oxalis acetosella*), show protec-tive leaf closure movements in high light intensities. This results from turgor changes in the pulvinar cells at the conjunction of the three leaflets, and since it is simply a response to light intensity and not light direction, it is termed *photonasty* (Pfeffer 1906). A more subtle effect of light on leaf turgor movement is seen in the phenomenon of *heliotropism*, in which the lamina is maintained at a particular angle in relation to the light source and thus follows the daily movement of the Sun across the sky. (Some writers quite appropriately refer to this as 'phototropism', a directional response to light, but it seems useful to have a term that distinguishes such turgor movements from other light-directed move-

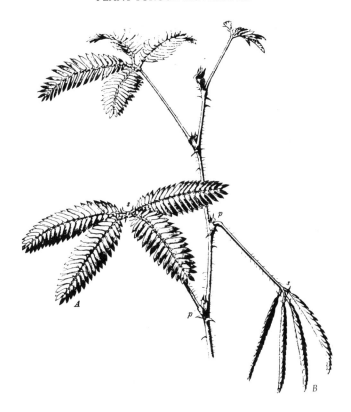

Figure 1.2 Thigmonastic movements in *Mimosa pudica* (the 'sensitive plant'). (A), Fully expanded leaf; (B), mechanically stimulated leaf; p, pulvini; s, secondary pulvini. (From Pfeffer 1906.)

ments that are based largely on growth.) Heliotropic leaf movements are generally of significance in maximizing photosynthesis, although in some plants the leaf blades are kept parallel to the Sun's rays in order to minimize the heat load (Smith 1984, Koller 1986, Prichard & Forseth 1988). Heliotropism is considered in more detail in Section 4.1.5.

Various forms of mechanical stimulation can initiate turgor movements in different kinds of plants, including such stimuli as contact (e.g. by insects), or slow and rapid flexure or bending (by wind or vibration). In the past, responses to these different stimuli have received different names, e.g. *thigmonasty* (response to contact) and *seismonasty* (response to flexure or vibration), but, as discussed in Chapter 5, such differences may be more apparent than real.

Mechanically stimulated turgor movements are generally associated with highly specialized forms of plant behaviour, none more so perhaps than the responses of the 'sensitive plant' (*Mimosa pudica*). This plant has

Table 1.2 Plant families in which different parts of the flower carry out turgor-based pollination movements (information from Simons 1981).

Floral part	Plant family
stamen filament	Berberidaceae, Cactaceae, Cistaceae, Compositae, Cucurbitaceae, Malvaceae, Portulaceae, Tiliaceae
stigma lobe	Bignoniaceae, Lentibulaceae, Liliaceae, Martyniaceae, Scrophulariaceae
style	Compositae, Mimulaceae
corolla tube	Gentianaceae
labellum	Orchidaceae
fused stamen-style	Stylidiaceae

doubly compound leaves with primary, secondary, and tertiary pulvini whose activities can completely change the form of the plant between the open and closed positions (Fig. 1.2). Superimposed on the normal nyctinastic cycle, the leaves close in response to a very wide variety of stimuli, including contact, vibration, extreme heat and cold, wounding, and toxic chemicals (Roblin 1979). The triggered movements occur relatively rapidly (within a second or so) and reopening takes place within 30–60 minutes. The extent to which closure spreads throughout the plant depends on the degree of stimulation or damage that has been inflicted. Presumably the adaptive advantage of this form of behaviour lies in protection against, say, rainstorms, high winds, or browsing insects. *Mimosa* is unusual in its extreme sensitivity, but plants such as clover and wood sorrel can also show some closure if the mechanical stimulation is severe enough.

Other highly specialized mechanically stimulated turgor movements are seen in insectivorous plants whose traps are sprung in response to contact (Pickard 1973, Hill & Findlay 1981, Simons 1981). Examples of these plants include the sundews (*Drosera* spp.), Venus flytrap (*Dionaea muscipula*), and the aquatic equivalent of the flytrap, *Aldrovandra vesiculata*. There are also the wholly submerged traps of *Ultricularia*, where the underwater bladders are maintained under a substantial negative pressure so that when a mechanostimulus triggers sudden loss of turgor of the retaining 'door' cells, the offending water insect is implosively sucked into the trap (Hill & Findlay 1981). Certain fungi also trap animals. When an unwary nematode enters the three-celled ring trap of *Dactylella* sp., any contact springs the trap and the nematode is caught by the sudden swelling of the trap cells (Jaffe 1985); in this case, the trapping mechanism unusually involves a sudden increase in turgor of the trap cells, which occurs in less than a second.

The pollination mechanisms of a very wide range of flowers provide many examples of mechanically stimulated turgor movements (Coombe & Bell 1965, Simons 1981). Virtually every type of floral organ seems capable of becoming adapted to produce a turgor movement that increases the chances of pollination (Table 1.2). The turgor movements in the flowers of the 'trigger plants' (Stylidiaceae) have received special study (Findlay 1984). In the flowers of these plants the reproductive organ is a stylidium, or fused stamen–style. In response to mechanical stimulation of the stylidium, extremely rapid changes in cell turgor cause the organ to flip through an arc of 300° and bang against the opposite petals within 25 milliseconds, a movement that must rank as one of the fastest in the plant world.

Some of these movements are discussed further in Section 5.3, in relation to mechanisms of reception and transduction of mechanostimuli by plant cells.

General comments

Turgor mechanisms enable plants to carry out rapid and repeatable movements, and are found in all three of the general biological 'problem areas' (Box 1.1): they are involved in protective roles (against high light intensities and mechanical damage); in processes of food collection, both in general (stomatal operation) and in specialized situations (insect capture); and in reproduction (pollination mechanisms).

Many of these examples illustrate the point that, presumably because of its potential for providing rapid movement, the turgor mechanism seems to be of particular significance in situations in which there is a high degree of plant–animal interaction. In some ways, of course, the types of turgor movements in which contraction of a small group of cells is translated into the movement of an organ lever are directly analogous to the operation of animal skeletal muscle, where contraction of a particular cell type also brings about movement of a (bone) lever. Indeed, analysis of the force–velocity characteristics of movement at the primary pulvinus of *Mimosa* showed that values are comparable to the equivalent parameters in insect muscle (Balmer & Franks 1975). It has been suggested (Hill & Findlay 1981) that it was perhaps the presence of the cellulose wall in plant cells that played a large part in the emergence of the turgor mechanism, rather than contractile proteins, as the basis for rapid organ movement in the plant world.

11

Box 1.2   Investigation of plant movements

Most plant movements occur either too rapidly or too slowly to be directly observed by us, within the time scale imposed by our sense of vision. Nevertheless, the end results of many plant movements have been known since earliest times. Around 400 BC, Androsthenes, the archivist of Alexander the Great, recorded daily (nyctinastic) movements in leaves of tamarind, and Pliny noted the (nastic) closure of clover leaves in bad weather (Sachs 1890). Such sporadic observations continued throughout the Middle Ages and later. Caesalpino noted the (circumnutatory) movements of tendrils. Berelli in 1632 described the irritability to touch of the stamens of *Centaurea* sp. And Robert Hooke, perhaps better known for his microscopical observations of cork cells, recorded in 1667 the touch sensitivity of the leaves of the then recently imported *Mimosa*. (Historical aspects of tropic growth movements are discussed in Section 2.2.)

The scientific investigation of plant movements is considered to have begun in 1729 with De Marian's demonstration that the rhythmic (nyctinastic) movements of bean leaves continued even when the plants were maintained in continuous darkness. An early classification of plant movements was carried out in the 1750s by Linnaeus, who introduced the term 'sleep movement' and devised an actual 'flower clock', by noting the different times of opening and closing in the flowers of a range of species (the effects of temperature (Section 1.3.2) made it fairly inaccurate). However, plant movements were still considered rather as oddities until the studies of the great botanists of the 19th century. Sachs, Haberlandt, Pfeffer, and Charles Darwin were all greatly intrigued by the topic, and during this period the anatomical and much of the physiological background to plant movement was established. It was also at this time that a real awareness developed that plants were sentient organisms that reacted to their surroundings, and that movement was often a visible manifestation of such reaction.

Darwin, in particular, was fascinated by plant movement, and he not only studied such obvious movements as occur in the tendrils of climbing plants and the trapping organs of insectivorous plants (Darwin 1875a, b), but he also characterized the constant nutatory movements of normal growth and the tropic and nastic movements of directional growth (Darwin 1880). Darwin was also among the first to actually record the process of movement, using a simple but ingenious method that involved attaching a small marker to the tip of the organ of interest, and placing the plant in front of a white card upon which was a black dot (Darwin 1880). The whole assembly was then viewed through a glass plate, and the course of organ movement recorded by marking a dot on the glass at the point where the marker and the background dot were viewed in alignment; the pattern of movement over a period of time could be shown by joining up the dots on the viewing plate (see Fig. 1.3a).

Nowadays, of course, modern techniques and equipment go some way towards substituting for the patience and insight of Darwin, and time-lapse photography brings plant movements into our perceived time scale of events. Much of the recent research interest into tropisms, for example, has resulted from the application of these 'new' techniques to the investigation of 'old' phenomena. Several laboratories have analysed the actual growth events that occur during tropic curvatures, rather than simply measuring the extent of curvature, and this has given new insight into the nature of tropic responses (see Ch. 3, 4). Different forms of equipment have been used but the basic technique utilizes some kind of time-lapse record of movement in suitably marked plants. Subsequent frame-by-frame measurements of the rates of displacement of the marks gives information about the changing patterns of growth in different zones of the responding organ (Digby & Firn 1979, Hart *et al.* 1982, Iino & Briggs 1984). The use of video equipment rather than cine film to record growth offers several more advantages. It not only avoids the delay of film processing, but also, since some types of video camera are sensitive to wavelengths of greater than 800 nm, it allows plant behaviour to be indirectly observed under conditions of complete 'plant darkness'. Video techniques also lend themselves to semiautomatic data processing and growth analysis (Jaffe *et al.* 1985, Berg *et al.* 1986).

1.3   PLANT GROWTH MOVEMENTS

1.3.1   *General features of plant growth*

There is movement in all plants, of course, as organs grow, expand, and generally fill up the space around themselves, and, since plant growth is much less finite than that of most animals, this enables a plant to cover considerable distances both above and below the ground. Often this form of movement over distance is carried out by specialized organs, such as stolons, runners, and so forth, and if these organs continually dieback in the older positions, then the plant literally does 'move'.

Again, in very many species the subterranean overwintering organs, such as bulbs, corms, tubers, etc., are physically pulled downwards into the soil by so-called contractile roots (Fahn 1982). This means of avoiding frost damage is common in many dicotyledonous (dicot) species, such as *Taraxacum*, *Daucus*, *Trifolium*, *Oxalis*, and *Medicago*, and in many bulbous or cormous monocotyledonous species (monocots), such as *Narcissus* and *Hyacinthus*. (The 'contraction' is generally through the actual collapse of horizontal rows of cells within the specially developed fleshy roots (Fahn 1982). The environmental signals that induce the development of such roots include exposure to wide day–night temperature fluctuations or light (Halevy 1986), either of which conditions would indicate a disadvantageous position too near the soil surface.)

However, plants can also alter the orientations of their various parts by means of differential growth between or within different organs. This ability to change their postures by altering their growth patterns is unique to plants, and derives from particular characteristics that distinguish the growth of plants from that of most animals. These relate to:

(a) *Spatial aspects*. Growth in a plant is localized largely within particular regions; growth in length takes place from meristems that are usually located at the ends of organs, and growth in thickness occurs from intercalary meristems.
(b) *Temporal aspects*. A plant retains the ability to grow throughout most of its life, and many tissues and organs often show renewed growth after considerable periods of quiescence.
(c) *Environmental aspects*. A major distinctive feature of plant growth is the extent to which it is influenced and controlled by environmental factors; the environment not only determines the eventual size of the plant, but also regulates many (most?) developmental aspects of growth, including the cessation of growth, the initiation of growth, and the establishment of differential patterns of growth.

All of these features interact to make plant growth much more malleable than that of most animals, and it is this malleability that allows growth to be used by plants as a means of continually adjusting the orientations of their various parts throughout most of their lives.

1.3.2   *Types of growth movements*

Nutational movements
The tips of most growing organs, including those of roots, shoots, and coleoptiles, move laterally through arcs or spirals as they grow, rather than proceeding along a straight line. These are called nutational movements, and they can range from the fairly irregular 'nodding' movements of very young seedlings to the much more regular rotatory movements seen in the *circumnutation* of older organs (Fig. 1.3). The adaptive significance of these movements is fairly obvious in the wide, so-called 'searching' nutations of twining stems and tendrils, but is not so clear in other organs. (Darwin considered that they aided passage through soil and litter.) However, their ubiquitous presence means that they must be taken into account in any analysis of other growth movements such as tropisms (Baskin 1986).

The form of circumnutation varies according to the species, from those with a more or less circular movement to those with elliptical or even pendulum-like movements (Johnsson 1979). The direction of movement also varies, e.g. clockwise in hops (*Humulus lupulus*), anticlockwise in bean (*Phaseolus vulgaris*). The speed and extent of movement varies not

(a)

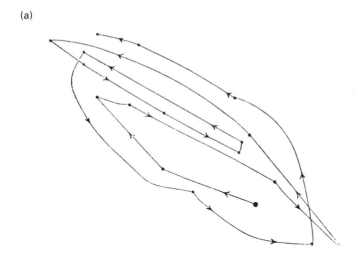

(b)

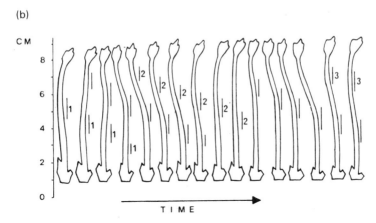

Figure 1.3 Circumnutatory movements. (a) Pattern of movement in a seedling of *Brassica oleracea* (turnip) over 10 h 45 min; the movement of the hypocotyl tip was recorded at intervals by viewing vertically downwards (see Box 1.2). (From Darwin 1880.) (b) Silhouettes of the first internode of *Phaseolus multiflorus* (runner bean) at intervals of 15 min; vertical bars trace the course of waves passing down the stem (from Johnsson & Heathcote 1973).

only between species, but also within a single individual according to the stage of development and growth conditions, particularly temperature. However, a single cycle of movement is generally completed within an hour or so. The two parameters that are used to formally describe a pattern of nutation are its 'period', or time for one complete cycle, and its

15

'amplitude', or range of movement from the central position to the extreme position.

Circumnutation derives from regular changes in the rates of cell elongation along an organ (Fig. 1.3b). It is generally held that waves of differential growth move basipetally down a circumnutating shoot (Johnsson 1979), although it has been shown recently that rhythmic fluctuations in cell turgor probably are involved as well, at least in circumnutation of tendrils (Millet *et al.* 1987). Various models of circum-nutation have been proposed (all discussed in Johnsson 1979) which suggest that the underlying fluctuations in growth rate are either caused by external stimulation or result from the effects of some endogenous factor.

The possibility of external regulation by gravity has received particular consideration. That is, several models suggest that circumnutation results from continually changing gravitational stimulation as the growing organ overshoots the vertical position, recovers and overshoots in the opposite direction, and so on. However, several arguments can be put forward against this:

(a) During circumnutation, the organ rarely, if ever, passes through the vertical position, i.e. if the organ is 'hunting' for its position of gravitational equilibrium it never actually finds it, even temporarily.

(b) Nutatory movements continue during a gravitropic response (Heathcote 1982), suggesting that the two growth movements are separate phenomena. [However, a contrary view has been expressed (Ney & Pilet 1981)]

(c) There is lack of correlation between nutation and gravitropism in several respects, including responses to changes in organ orienta-tion, to treatments with red light and with active transport inhibi-tors, and to excision of the apical bud (Britz & Galston 1982), again suggesting that they are separate phenomena.

(d) Exposure of organs to deliberate pulses of transverse gravitational stimuli has no effect on the nutational period (Britz & Galston 1982). (Such a treatment can affect the amplitude of circumnutation, however, and can also often synchronize the phase of circumnutation in different individuals.)

(e) Finally, circumnutation has been observed in organs exposed to minimal gravitational stimulation, both during Earth orbit (Brown & Chapman 1984) and in a simulated zero-*g* situation on a klinostat (Brown & Chapman 1988).

It would therefore appear that some form of endogenous regulation is responsible for circumnutatory growth patterns. Suggestions in this area (also reviewed in Johnsson 1979) involve several possibilities. For

16

example, oscillatory phases of growth at the cellular level, say between phases of expansion and consolidation, may be amplified into periodicities of growth at the organ level. Alternatively, there may be fluctuations in some aspect of growth regulator activity. It is unlikely that fluctuations in production of regulator are involved, since a decapitated organ provided with a steady supply of auxin still circumnutates, but there are known to be fluctuations in the rate of transport of auxin of the order of 20–30 minutes (see Koukkari & Warde 1985). However, an intriguing possibility is that endogenously generated mechanical strains may regulate growth rates. For example, in both the sporangiophore of *Phycomyces* (Dennison 1979) and the peduncle of dandelion (Clifford *et al.* 1982) it has been shown that compression tends to stimulate the growth rate, whereas tension inhibits it. Therefore, it may be that in a growing organ natural differences in tissue growth rates generate a certain amount of sporadic movement. These movements would themselves generate tensions and compressions on opposite sides of the organ, which could establish self-sustaining rhythmic changes of growth rate. (In such a situation gravity would not be the initiating or controlling factor, but would influence the amplitude of the movement.) It is noteworthy in this respect that time-lapse films of the development of circumnutation in dicot seedlings invariably show initial irregular nodding movements in very young seedlings; only when the seedlings reach a minimum height of around 2 cm do rhythmic circumnutatory movements become apparent.

The results of tissue tensions and compressions may be to change the properties of the cell membranes (Edwards & Pickard 1987), with consequent effects on either the transport of a growth regulator or the sensitivities of the cells towards growth regulator. These aspects are discussed further in Chapter 3, with regard to gravitropic responses, and in Chapter 5, with regard to growth responses to other forms of mechanical stimulation.

Nastic movements

Nastic movements are due to either a single wave or repeated patterns of differential growth which result in a change in orientation or shape of an organ. The differential growth can be part of the normal developmental pathway of the organ, or it can be initiated directly in response to some environmental stimulus. However, in both situations the actual pattern of differential growth, and thus direction of movement, is based upon some endogenously determined asymmetry within the responding tissues. In environmentally induced nastic movements, therefore, the actual stimulus is often fairly diffuse without any strong directional component.

The term *epinasty* is used in its widest sense (Palmer 1985) to describe any situation in which the upper side of an organ grows at a faster rate than the lower side. (In the opposite, and less common, case of hyponasty, the lower side grows at a faster rate.) Epinasty occurs during normal leaf development when the young vertically oriented leaf moves to its mature horizontal orientation as a result of a wave of differential growth moving basipetally along either the leaf petiole (dicots) or the leaf itself (many monocots). However, the regulatory mechanisms responsible for this epinastic development are not at all clear. The growth regulators auxin and ethylene are obviously involved (Kang 1979), and several models have been proposed that rely on various forms of feedback interaction between ethylene and auxin action or auxin transport (reviewed in Palmer 1985). However, epinasty does not seem to be due simply to a transverse concentration gradient of growth regulator and it is thought that some sort of differential sensitivity of the tissues must be involved (Palmer 1985).

A wide range of exogenous factors can bring about such extreme epinastic development that the petioles become twisted and distorted. These factors include flooding, injury, disease, herbicides, and pollutants. The common feature in these effects is the stress-induced production of ethylene, which then provokes the epinastic growth (Palmer 1985). However, again in these cases, the exogenous factors and ethylene are thought to be acting upon a pre-existing physiological asymmetry within the tissues, rather than creating such an asymmetry.

Gravity is another exogenous factor which influences the extent to which epinasty is expressed. For example, in petioles and lateral branches, epinasty acts to bring the organ down to a horizontal orientation, whereas negative gravitropism acts to raise the organ to the vertical (Kaldeway 1962). The interaction between epinasty and gravitropism in the orientation of lateral branches is discussed further in Chapter 3.

Epinastic growth also occurs in the plumule of dicot seedlings, which emerges through the soil with the apex in the form of an arch or hook. (This serves to protect the delicate apical meristem as it is pulled rather than pushed up through the soil). The form of the hook is maintained by a complex pattern of differential growth, involving faster growth on the outer convex side at the top of the hook, and also on the inner concave side at the bottom of the hook, i.e. there is a reversal of the transverse growth gradient within the hook region (Silk 1984). During upward growth the hook remains a morphological entity even though its constituent cells change as they elongate and pass out of the region. In fact, the hook has been described as a 'standing epinastic wave riding the expansion zone of the plumule' (Palmer 1985). Maintenance of the differential growth rates across the hook is generally considered to be due to ethylene (Palmer 1985): the hook region is the major site of ethylene production in the seedling; in some species ethylene brings about hook closure; and when hook opening is induced by light, ethylene production ceases. Again, ethylene seems to be acting on some endogenously regulated pattern of differential tissue sensitivity.

One case of epinastic leaf growth in which ethylene does not seem to play a role is in young cereal leaves (Kang 1979). These leaves emerge in a longitudinally rolled condition, and, under the influence of red light, unroll through a greater expansion rate in the cells on the inner adaxial side. Light is thought to bring about these effects by inducing the synthesis and release of gibberellin from the cell plastids.

In many species, e.g. poppy, fritillary, daffodils, the orientations of the flower heads change during development through changes in the patterns of differential growth in the flower stalk. (These movements are of considerable adaptive significance, with a downwards orientation of the bud stage conferring protection, an upright position of the open flower enhancing the chances of pollination, and a final 'drooping' position of the fertilized flower aiding seed dispersal.) In many cases, these changes in growth pattern are brought about by changes in responses to gravity (Ch. 3). In others, however, they are due to endogenously regulated patterns of epinastic growth (MacDonald *et al.* 1987). For example, in some species of cyclamen (*Cyclamen hederifolium*) the events of fertilization initiate a resumption of growth in the flower stalk in such an extreme epinastic pattern that the stalk coils downwards over a period of five days or so (Fig. 1.4), until eventually the seed pod is actually buried in the surface litter. The differential growth of coiling is thought to be due to an increased production of auxin consequent upon fertilization, but exactly how this auxin brings about differential growth, whether by differential transport or by a predetermined differential sensitivity of the tissues, is not known.

Cases in which exogenous factors have a more direct and immediate influence on nastic floral movements are seen in the effects of light and temperature on the opening and closing movements of many flowers, movements of obvious adaptive significance in protecting the delicate reproductive organs. The mechanisms responsible for these movements are not as clearly delineated as those underlying leaf nyctinasty, but they generally seem to involve some growth (Pfeffer 1906). The flower movements also seem to be much less strictly under the control of circadian timing mechanisms. An increase in light intensity induces opening in many species, but effects of temperature often overrule this photonastic response. For example, flowers of daisy (*Bellis perennis*), dandelion (*Taraxacum officinale*), and hawkbit (*Leontodon* spp.) do not open even on bright sunny days if the temperature is below about 5 °C (Pfeffer 1906), thus maintaining the protection of the reproductive organs when there are unlikely to be any insect pollinators around. Examples in which temperature effects are wholly responsible for the opening and

19

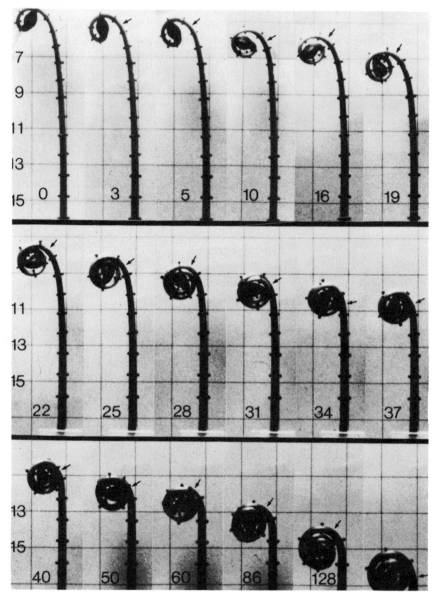

Figure 1.4 Epinastic growth movements in the flower stalk of *Cyclamen hederifolium*; the stalk began coiling at 0 h, and thereafter photographs were taken at 3 h, 5 h, etc.; marker beads allow growth rates to be measured in different regions, and arrows indicate the moving region of greatest growth rate (from MacDonald *et al.* 1987a).

closing movements are seen in the *thermonastic* responses of spring flowers such as *Crocus* and *Tulipa*. Here the movements are due to the differential effects of temperature on the growth rates of the inner and outer epidermis of the petals (Crombie 1962). The cells of the inner epidermis have a higher temperature optimum for growth than those of the outer epidermis. Therefore, a rise in temperature causes the inner epidermis to grow faster than the outer, and the flower opens; a fall in temperature causes the outer epidermis to grow faster than the inner, and the flower closes. The growth responses are extremely sensitive, with opening or closing movements being initiated by as little as 0.5 °C change in air temperature.

Thigmonastic responses to contact stimuli occur in many climbing tendrils, and are discussed along with thigmotropic responses in Chapter 5.

Tropic movements

In a tropic response differential growth within an organ is not only initiated by the action of some environmental stimulus, but the actual pattern or distribution of growth is also regulated by the stimulus. The stimulus itself generally has a strong directional component, and the change in orientation of the organ or the direction of movement therefore bears some relationship to the direction of the stimulus.

The major types of stimuli that induce tropic responses are gravity and light. Mechanical stimuli can also induce directional growth movements, particularly in roots and in climbing tendrils. And other types of stimuli, such as chemicals, injury, and electrical signals, can exert significant directional effects on the behaviour of particular organs or in particular situations. All of these different types of tropic stimuli act through the regulatory sequence of stimulus reception, physiological mediation, and induction of the growth response (Ch. 2).

The nature of the growth responses that underlie the tropic change in orientation of an organ differ markedly between different types of plants and different types of tropisms (Ch. 2). They can involve the change in position of a growing point, or change in the region of growth. They can involve growth stimulation, or growth inhibition, or both. And the regulatory mechanisms by which these changes in growth are effected also differ. Even within higher plants, the traditional view that all tropic responses result from differential distributions of the growth regulator, auxin, is subject to serious challenge. In some tropisms at least, the regulatory involvement of changes in the cell membrane and changes in electrical activity are now recognized. These aspects, too, are all developed further in subsequent chapters.

General comments

Tropisms provide a striking example of the extremely close and malleable relationship that exists between plant growth and the environment. Tropic movements are the major means by which a plant adjusts itself and its parts in direct relation to its surroundings. They are therefore of prime importance in enabling the plant to optimize its harvest of environmental resources, i.e. in enabling it to continually move, adjust and orient its organs most favourably for the uptake of energy, water and nutrients. The extent to which a plant requires this capacity to orient its organs obviously varies according to its stage of development and its situation. For example, it is crucial that young seedlings with limited food reserves reach external supplies of energy and nutrient as quickly as possible, and tropic responses are strongly evident at this stage. However, tropisms are also important in mature plants, inasmuch as they enable the plant to adjust to a change in its environment, or even to react to some environmental catastrophe. For example, if a plant falls over, it 'gets up' by tropic growth; if a plant part becomes shaded, it 'moves away' by tropic growth. Again, the function of an organ may change during development. The direction of growth required of a flower stalk, for example, is not necessarily the direction that is of greatest advantage when that same stalk eventually bears seeds.

Tropisms are thus not only significant factors in the determination of the whole form and bearing of a plant, they are also a major means by which a plant asserts itself and gently inserts itself into its environment.

CHAPTER TWO

Introduction to tropisms

2.1 GENERAL DESCRIPTION

2.1.1 *Types and terminology*

A tropism is a directional response of a plant organ to a directional stimulus in the environment, usually through some form of differential growth. The definitive characteristic is that the pattern of growth and direction of response bear some strong relationship to the direction of the stimulus.

Various terms are used to describe the direction of response. If the stimulated organ becomes oriented in the same plane as the stimulus, it is described as an orthotropic, or more rarely, a parallelotropic, response, and is said to be positive if it is towards the source of the stimulus, and negative if it is away from the stimulus. The responses of seedlings are usually orthotropic and wholly positive or negative, but those of more mature tissues show greater variation. If the organ becomes oriented at an angle to the direction of stimulation, as in lateral branches and some roots, the response is described as being positively or negatively plagiotropic. And if the orientation is at right angles to the stimulus, it is termed diatropic.

Tropisms are named fairly obviously according to the types of environmental factor that initiate them, the major ones being gravitropism (or geotropism) and phototropism. In general the main axis of a plant is orthogravitropic, with negatively orthotropic stems and positive roots; lateral branches usually show some degree of negative plagiogravitropism, and secondary roots are often positively plagiogravitropic; rhizomes and other horizontally growing organs are diagravitropic. However, there are many exceptions to these generalities (see Ch. 3).

The aerial parts of all plants generally show some form of positive phototropism, whereas only some types of roots and subterranean organs are negatively phototropic. The term phototropism strictly refers to the directional growth of an organ in response to blue light. However, light is of such crucial importance for plant growth and development that

23

there are, in fact, several additional means by which it can influence the orientation of plant organs. These include heliotropism, or sun-tracking, which is shown by many leaves and flowers and which often involves changes in cell turgor rather than growth. They also include red-light-induced 'shade effects' on the growth of higher plants. And red and blue light also have strong positive and diatropic effects on the sporophyte and gametophyte generations of lower plants. These aspects are discussed further in Chapter 4.

Directional responses to physical contact or touch, thigmotropism (in the older literature 'haptotropism'), play an important role in the orientation of many plant organs (Ch. 5). These are most apparent in certain specialized organs, such as the tendrils of some climbing plants and the trapping organs of insectivorous species. However, over 100 years ago Charles Darwin pointed out that this type of response is also generally present in roots. He maintained that roots are negatively thigmotropic (bend away from contact) in the region immediately behind the tip but show a positive response (bend towards contact) in regions further from the tip, and suggested that such behaviour helps the root to wriggle through the soil while maintaining nutritional contact with soil particles.

Another tropism that may be of widespread occurrence, and which is also particularly evident in roots, is a directional response to injury, traumatropism. When injured, roots bend away from the side that is injured. Stems, on the other hand, curve towards the injured side. These wound responses are not due simply to mechanical damage or to cell shrinkage, say by the loss of water, but involve definite growth reactions in regions some distance from the site of injury (see Ch. 6).

Other tropisms are known, but in some cases these are not common to all types of plants and in others their significance to general plant development is not clear. For example, directional responses to chemical stimuli, chemotropism, are important aids to sexual reproduction in motile lower plants and in many fungi. But in higher plants directional growth responses to chemical factors seem to be limited, perhaps, to the growth of the pollen tube down the floral style towards the ovary, and to the behaviour of trapping organs in some insectivorous plants (see Ch. 6). Again, directional responses of the roots of some plants towards water, hydrotropism, have been described, but the general occurrence of this phenomenon under field conditions has not really been established (see Ch. 6). And directional responses to linearly polarized light, polarotropism, have been described in simple organs such as algal cells and protonemal filaments of mosses (see Ch. 4), but the significance of this phenomenon to the directional growth of multicellular organs in natural light is not clear. The pattern and direction of growth in many plant organs can also be influenced by electric and magnetic fields, but again the significance of such responses to general plant development is not clear.

In the older literature even more responses have been named as tropisms, but do not seem to be reliably established. For example, 'thermotropic' responses of seedlings have occasionally been noted (Aletsee 1962a), but usually the studies have not been carried out under controlled conditions and in some cases the actual nature of the causal factor as being heat, rather than, say, loss of water, has not been confirmed. And directional growth responses of some fungal hyphae seemingly in response to the direction of water flow, 'rheotropism', may in fact be due to the presence of chemical factors in the flowing water (see Ch. 6).

There is one further important response or form of behaviour that occurs in all tropisms. The term 'autotropism' is used to describe the straightening reaction that must take place (if it did not, a tropically responding organ would eventually form a loop). At first glance, this straightening of tropically curving organs may seem to be due simply to some sort of equalization of the tropic stimulus as the organ achieves an equilibrium position. However, there are several reasons for this being an unlikely explanation of autotropism. For example, in the case of gravitropic curvature, the straightening response continues to occur after the plant has been placed on a rotating klinostat (i.e. after gravitational influences have been experimentally equalized). Secondly, the growth events that underlie the straightening response of a gravistimulated organ are actually initiated well before the organ becomes vertical, and in some cases in regions of the organ that never become vertical (Firn & Digby 1979). And, finally, autotropic straightening occurs not just in gravitropism but in all tropic responses, including thigmotropic tendril coils if the contact stimulus is removed [its apparent absence in some cases of coleoptile phototropism (Pickard 1985b, Nick & Schäfer 1988), but its presence in others (Firn 1986a), may be due to differences of definition]. Thus although autotropism is based upon distinct and specific growth reactions, it is unlike other tropisms in that it is not a response to an external directional stimulus. It may be that the autotropic 'stimulus' is simply a mechanical effect that results from tissue strains within a curved organ, which stimulate appropriate growth reactions until they are balanced out (see also the discussions on the possible growth regulatory effects of tissue tensions and compressions in relation to circumnutation, Section 1.3.2; gravitropism, Section 3.2.1; and responses to mechanostimuli, Section 5.3.2). However, whatever their origins, autotropic growth reactions are significant components of every tropic response.

(The term 'autotropism' is also used in a slightly different sense to describe the directional growth responses of fungal hyphae to the presence of other hyphae of the same individual or the same species, see Ch. 6.)

2.1.2 Stages in the tropic response

Like any other environmentally regulated phenomenon, a tropism can be regarded as a catenary system, i.e. as a chain of events which are causally linked. The stages involved can be somewhat arbitrarily categorized as:

stimulus reception → physiological mediation → biological response.

Such a sequential representation may be somewhat misleading since the system really involves not distinct and separate stages, but continuous and overlapping processes. For example, usually stimulation is still being received while other events, including the final responses, have already been induced, and in fact 'the temporal interweaving of the several aspects of (the system) forms a tight mesh' (Pickard 1985a). Nevertheless, this traditional sequential representation is still useful for purposes of description and discussion.

Stimulus reception

This stage consists of the action of a particular environmental factor upon a biological receptor, as a result of which the receptor is altered in some way. Implicit in the concept of a receptor being altered by the energy form of a particular type of stimulus is the principle that there must be appropriate types of receptors for the different types of environmental stimuli, i.e. there must be gravireceptors, photoreceptors, mechanoreceptors, and so forth. However, unlike the sensory receptors in most animals, plant receptors are not specialized cells or organs. Instead they are structures, or even simply molecules, that are possibly associated with some part of the cell membrane system, either on the cell surface or within the cell itself (Bentrup 1979). The actual identity of the receptor for any environmentally stimulated movement in higher plants has not been unequivocally established yet, although, because of the characteristic specificity of light absorption in relation to its action, something is known about the general type of pigment molecule that must act as the phototropic receptor.

In their sensory interactions with environmental stimuli, organisms must deal with what has been called the 'sensory paradox' (Hertel 1980). That is, the sensory systems must be able to respond over extremely large ranges of stimulus intensity (e.g. around six orders of magnitude in the light environment), and yet they must be able to detect very small differences in levels of stimulation. The solution to this paradox often seems to involve the phenomenon of sensory adaptation, in which responsiveness or sensitivity to a certain level of stimulus becomes less marked with continued exposure to that level of stimulus (a very hot bath rapidly becomes less uncomfortable without the temperature of the bath changing to any great extent). The mechanisms responsible for such adaptation are not always clear, and may be due to changes in the receptors themselves or in some other part of the receptor–response system. Nor is it clear to what extent this phenomenon is generally involved in the tropic responses of higher plants. However, it is definitely a feature of the phototropic behaviour of the *Phycomyces* sporangiophore, which can respond over a range of light intensities covering nine orders of magnitude (Dennison 1979, Hertel 1980).

Another aspect of stimulus reception that is of particular importance in directional responses is the means by which the direction or gradient of stimulus is detected. There are two quite different methods of achieving this, known respectively as 'temporal sensing' and 'spatial sensing'. In temporal sensing the organism compares the intensity of the stimulus at one point in time with its intensity at another time, and according to any change in intensity an appropriate behavioural response is induced. This method operates in organisms that are not large enough to be able to detect a significant difference in stimulus level within the confines of their own body size. The mechanism has been particularly well studied in relation to chemosensing in bacteria (McNab 1979), where a change in concentration of, say, a chemoattractant, induces a transient change in the pattern of flagellar activity and thus a change in the direction of motion. This form of response is discussed further in Section 6.1. Temporal sensing mechanisms are also involved in controlling the responses of ciliated organisms towards chemical and mechanical stimuli (Taylor & Panasenko 1984). And it is basically a form of temporal sensing that is reponsible for the light-directed motion, or phototaxis, of algae such as *Euglena* (Carlile 1980). In this organism, flagellar activity is controlled by a pigment complex which consists of a photoreceptor and a screening pigment, the eyespot. The eyespot casts a shadow on the photoreceptor and thus flagellar activity is temporally regulated according to the constantly changing relationship between the moving cell and the direction of light (Häder 1979, Feinleib 1980).

In spatial sensing mechanisms the intensity of the stimulus in one region of the cell, or organ, is compared to the intensity in another region, and an appropriate response is induced. This method of gradient detection is obviously more suitable for larger or more slow-moving organisms and is assumed tacitly to be operating in tropisms. The effectiveness of the method can be increased by several mechanisms which enhance the differential spatial sensitivity of stimulus reception. These may include differential location of receptors within the cell or organ, or some kind of differential receptivity as in, for example, a dichroic photoreceptor which preferentially absorbs light vibrating in a particular plane. Greater differential spatial sensitivity may also be achieved by methods that increase the gradient of the stimulus within the organ, e.g. by shading or screening effects. And in some cases the stimulus may create some sort of physiological gradient within the tissues, e.g. this may be involved in tropic responses to gravity (Wilkins 1984), where there is no significant difference in gravitational force between the upper and lower sides of a gravistimulated horizontal organ.

Physiological mediation

This stage is the 'black box' that encompasses all the processes between stimulus reception and the observable tropic response. It starts with the production of some kind of metabolic signal by the altered receptor, i.e. there is *transduction* of the energy form of the stimulus into a biochemical form of energy. Since a limited level of stimulus usually produces a relatively large biological response, there must also be some *amplification* of the signal within this stage. Common amplification mechanisms for biological processes include enzyme action, the involvement of regulatory chemicals, including hormones and intracellular secondary messengers, and change in the properties of the cell membrane. Although plant growth regulators are involved in tropisms, the actual details of tropic signal amplification are not yet clear. [In animal sensory phenomena, considerable amplification occurs within the transduction stage, involving transient change in the ionic permeability of the membrane of the sensory cell. In the vertebrate eye, for example (Eckert & Randall 1983), absorption of one photon of red light with an energy of 3×10^{-19} J generates a receptor current containing about 5×10^{-14} J of electrical energy, an amplification of about 1.7×10^5 times.]

This stage may also involve *transmission* of the transformed signal from the site where the stimulus is received to the region where the tropic growth response takes place. There has been some recent controversy over whether this actually occurs in tropisms (Firn & Digby 1980, Trewavas 1981). Certainly, the extent of spatial distinction between regions of tropic sensitivity and regions of growth response is not as great for all tropisms as originally believed, particularly for those in aerial organs of higher plants. However, in many organs and responses some form of signal transmission is obviously involved.

It is traditionally held that in plants information is conveyed from one region to another by chemical means, i.e. by plant hormones. This concept originated from the studies of tropisms by Charles Darwin, continued through what have come to be viewed as classic experiments by several investigators (see Section 2.2), and culminated in the isolation of the growth regulator, auxin. And since then, of course, other growth-regulating chemicals have also been identified. However, although such chemicals certainly influence growth, there is now some controversy about whether they truly act as hormones, i.e. about whether their actual movement conveys growth regulatory instructions throughout the plant (Firn 1986b, Trewavas 1986, Weyers *et al.* 1987).

There is another, perhaps less familiar, means by which information is transmitted within the plant. It is now apparent that electrical signals play a highly significant role in initiating and co-ordinating plant responses (Pickard 1973, Bentrup 1979, Simons 1981, Frachisse *et al.* 1985, Davies 1987). In sensory phenomena in animals, the major effect of most forms of external stimuli is to generate an *action potential*, or transient change in the transmembrane potential of the receptor cell. This often involves a temporary depolarization, i.e. a temporary loss of membrane potential due to inward flow of current, and derives from a transient change in membrane permeability towards ions. In some cases, the inducing stimulus can be at a sub-threshold level and generates what is called a receptor potential; accumulation of enough receptor potential can trigger off an action potential. The crucial characteristic of an action potential is its self-generating or regenerative nature, i.e. an initial change in membrane permeability, or number of opened 'ion gates', causes more ion gates to open and the effect builds up until it spills over into the generation of the action potential. It is this regenerative property that enables the action potential to be propagated through tissues until it eventually initiates some biological response in an appropriate motor cell. (The term action potential was originally used to describe an electrical signal that brought about some biological action, but now it is routinely applied to a temporary, regenerative change in transmembrane potential.)

The occurrence of action potentials in plants is particularly well documented in relation to the specially sensitive tissues of *Mimosa pudica* and the insectivorous plants (Pickard 1973, Simons 1981) where rapid turgor movements are triggered by mechanical stimulation. [In fact, electrical activity in plants was first described in the Venus flytrap by Burdon-Sanderson in 1873 (Pickard 1973), i.e. long before the discovery of chemical regulation of plant activities.] However, they have also been detected in less specialized tissues in response to various forms of stimuli, including contact, wounding, light, chilling, and reorientation in relation to gravity (Bentrup 1979, Frachisse *et al.* 1985, Davies 1987, Edwards & Pickard 1987). They are transmitted at rates of 5–500 mm s^{-1} (cf. animal values: sea anemone 100 mm s^{-1}, cat 85 000–165 000 mm s^{-1}), and are generally based upon a transient depolarization involving efflux of K$^+$ and influx of Ca^{2+} (Davies 1987).

Plant action potentials have been studied largely in relation to turgor movements. However, it has been suggested that action potentials may also be involved in regulating changes in growth rate, either directly, through changed concentrations of regulatory calcium ions (Davies 1987), or indirectly through changes in membrane properties affecting levels of, or sensitivities to, growth regulators (Pickard 1973, Edwards & Pickard 1987). The regulatory involvement of electrical signals in plant responses is discussed in later chapters, particularly with regard to responses to mechanical stimuli (Ch. 5).

The cell membrane thus plays a central role in the physiological mediation of many external stimuli in plants. It is often the location of the receptor, and it is also often intimately involved in the transduction and amplification of the stimulus into a physiological signal, through changes in its permeability properties (Bentrup 1979). It must also be involved in any transmission of the signal as an action potential, and in the subsequent biological effects, certainly in the initiation of turgor changes in specialized motor cells but perhaps, too, in the regulation of growth rates. [The cell membrane also plays a central role in the reception and transduction of sensory stimuli in micro-organisms (Carlile 1980, Naitoh 1984, Taylor & Panasenko 1984). In *Tetrahymena* and *Paramecium*, chemostimuli and mechanostimuli both induce transient changes in the cell membrane which result in the influx of Ca^{2+} and the efflux of K^+. These transient changes in ion concentrations influence flagellar activity, to bring about changes in the organism's rate or direction of movement.]

Biological response

In any consideration of sensory responses, animal or plant, it is important to keep in mind that the stimulus only acts as an initiating factor. The nature of the response, and the energy which powers it, are determined by features within the cells or tissues themselves. This has two implications. First, it means that the same stimulus can provoke different responses in different tissues, or even in the same tissue at different stages of development. Secondly, it means that there is usually no straightforward linear relationship between the level of the stimulus and the level of response. (The situation is rather analogous to a gun going off: the speed of the bullet is not influenced by the force on the trigger.)

In all sensory phenomena, the relationship between stimulus intensity and extent of response is always some kind of power function, of the form:

$$\text{response} = k(\text{stimulus intensity})^x$$

where x is a constant exponent for any particular system (Shropshire 1979). In some cases, the response is proportional to the logarithm of stimulus intensity. It is not really clear what is responsible for this non-linear relationship, i.e. whether it is due to some feature of the receptor or to some other part of the mediation system. However, its biological significance is clear. It enables a sensory system to function over the huge range of stimulus intensities usually encountered in nature. For example, a log-linear relationship between stimulus intensity and response means that the high intensity end of the stimulus scale is 'compressed', thus greatly extending the range over which changes in stimulus level can be detected. Again, it means that a given percentage change in stimulus intensity invokes the same increment of change in response, e.g. a doubling of stimulus intensity at the low end of the range gives the same increase in response as a doubling at the high end. That is, change in stimulus intensity can be sensitively detected over the whole range of stimulus level. (This kind of relationship between stimulus intensity and response is similar to that governing subjectively perceived changes in stimulus intensity and is known in psychology as the Weber–Fechner Law, a term that is occasionally used in studies of plant sensory behaviour.)

Thus the nature of the relationship between stimulus intensity and response, together with the phenomenon of sensory adaptation and any physiological mechanisms for the enhancement of differential spatial sensitivity of stimulus reception, all contribute to the ability of a system to deal with the sensory paradox of having to detect small changes in stimulus intensity against background intensities that usually extend over many orders of magnitude. It is this ability to detect changes or differences in level of stimulus, rather than the absolute level of stimulus, that is of crucial importance in detecting any gradient of stimulus. And of course, by definition, tropisms are growth responses related to the direction, or gradient, of the stimulus.

2.1.3 Type of growth response

The actual nature of the growth response differs in different types of plants, but all tropic responses fall into one of two categories, 'bulging' or 'bowing' responses (Pohl & Russo 1984). The bulging type of response (Fig. 2.1a) is restricted to unicellular, tip-growing organs of lower plants, such as algal filaments, rhizoids, fern and moss protonemata, and fungal hyphae. It results from a displacement of the growing tip so that the filament simply bulges out in a new direction. Within this category, two sub-types of response are possible (Page 1968). In the more usual situation the growing point is displaced *towards* the stimulus, and this results in a *positive* tropic response (the side nearest the stimulus grows faster). Alternatively, and less commonly, the growing point can be displaced away from the stimulus to result in a negative tropic response.

The bowing type of response (Fig. 2.1b) results from the establishment of some pattern of differential growth so that the organ bends towards or

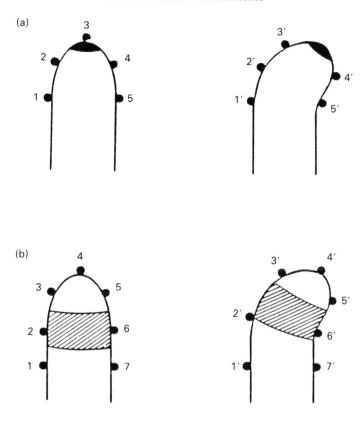

Figure 2.1 Different forms of tropic growth response; comparisons of the markers 1–7 and 1′–7′ allow the regions involved in the response to be determined. (a) Bulging response; change in the position of the apical growing point (shaded black). (b) Bowing response; differential growth in the sub-apical growing zone (hatched area). (From Pohl & Russo 1984.)

away from the stimulus. This type occurs also in unicells or filaments if the organ is elongating by intercalary growth rather than at a growing tip. However, note that in this case an acceleration of growth on the side *nearest* the stimulus brings about a *negative* tropic response, whereas a positive tropic response requires growth to be relatively slower on that side. (But just to make this situation really complicated, a light stimulus can be focused on to the far wall of a transparent, filamentous organ and bring about a positive tropic response through an acceleration of growth on the side *farthest* from the stimulus, see Section 4.1.2.)

It should be obvious already that, even in unicells, the various forms of tropic response can arise from markedly different types of growth reaction which must involve equally different types of regulatory mechanisms. The most effective way of distinguishing the type of growth that is involved in a response is by using some form of time-lapse technique to record the behaviour of surface markers (Fig. 2.1).

In multicellular organs, of course, a change in the direction of growth can only be achieved by some form of bowing response, and again there are several ways in which the pattern of differential growth can be established (Firn & Digby 1980, Firn 1986a). The possibilities consist of:

(a) stimulation of growth on one side of the organ;
(b) inhibition of growth on one side of the organ;
(c) differential growth stimulation across the organ;
(d) differential growth inhibition across the organ;
(e) stimulation of growth on one side and inhibition of growth on the other.

Again, it is necessary to know the particular pattern of differential growth in order to consider what types of regulatory mechanism may be operating.

2.1.4 Measurement of response

The eye is a poor judge of curvature (Silk 1984). For example, a short arc of a circle is perceived as being less curved than a longer arc of the same circle (Fig. 2.2). Equally, a simple measurement of degrees of curvature is entirely inappropriate as an objective and quantitative measure of a tropic response. In the first place, differences in amount of curvature between organs may simply be due to differences in the growth rates or the thicknesses of the organs (Badham 1984). And secondly, as we have already seen, similar forms of curvature can arise from markedly different patterns of differential growth. Therefore, to characterize and quantify a tropic curvature it should be the growth responses themselves that are measured.

In any investigation of growth rates, great care must be exercised in regard to the choice of temporal and spatial parameters of measurement (Silk 1984). Even in the seemingly simple experiment of placing marks along a plant axis in order to find out where growth is occurring, mistakes of interpretation can occur if the time intervals over which the marks are measured are too long (Silk 1984). This is because the marks themselves are being displaced throughout the period of measurement and the effects of growth are steadily accumulating over time; with longer time intervals it therefore becomes more and more difficult to accurately locate the regions of growth within these steadily lengthening zones. (The classical root-marking experiments of Sachs in the 19th century were mistaken in locating the region of maximum growth rate to

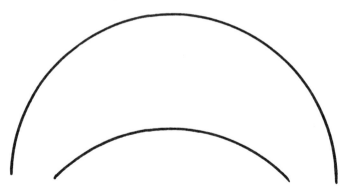

Figure 2.2 The eye is a poor judge of curvature; both curves are arcs of equal curvature (from Silk 1984).

the second millimetre behind the root tip; it is, in fact, a good 40 cell lengths farther back than this, see Erickson & Silk 1980.) This difficulty is overcome by choosing shorter time intervals of measurement or, even better, by using some form of relative elemental growth rate (e.g. mm mm^{-1} h^{-1} or percent h^{-1}) which gives the instantaneous growth rates of individual points at particular times.

The choice of spatial parameters of measurement is equally important. Within a tropic response, information is required about the location of curvature, the shape of the curve, and the contributions different regions make to the curve. For example, in a recent analysis of the gravitropic response of roots it was noted that curvature was not uniform, but consisted of two areas of curvature separated by a flatter region (Selker & Sievers 1987). These different contributions to curvature made by different regions of the organ must eventually be accounted for in regulatory models of root gravitropism.

There is another important, but often unrealized, feature of the growth events during curvature of an organ. This is that, for the organ to stop curving, there must be an actual *reversal* of the gradient of differential growth across the organ (Silk 1984); if the growth rates simply become equalized on each side, the organ would continue to curve in a circle. (The situation is analogous to that in the plumular hook, Section 1.3.2, where the growth gradient must reverse between the apical and basal ends of the hook.) Reversal of the cross-sectional growth gradient has been observed in gravitropically curving roots (Selker & Sievers 1987). Again, regulatory models must take account of these observed complexities of growth, and any putative agents of physiological mediation should show the same reversal of sign as the growth gradient.

2.2   HISTORICAL OVERVIEW

A better appreciation and understanding of the present situation in a subject is often gained from a greater awareness of its past. Therefore a brief outline is given here of historical developments in the investigation of tropisms. The reference sources are Sachs (1890), Reynolds-Green (1909), Went (1935), Went & Thimann (1937), Larsen (1962a), Went (1974), Heslop-Harrison (1980), Pohl & Russo (1984).

2.2.1   Early ideas

The general effects of light and gravity on the form and orientation of plants must have been fairly obvious, of course, from earliest times, but realization of the actual roles of these factors did not develop so obviously. For example, the Roman scholar Varro, writing around 100 BC, described the curvature of flower stems in light, but the English botanist Ray, in 1693, attributed this and other plant curvatures in sunlight to temperature differences on opposite sides of the organs. However, in 1758 DuHamel correctly asserted that such curvatures were due to light itself. Again, in 1703 Dodart noted the vertical orientations of shoots and roots and their generally linear relationship to each other, but suggested that these features were due to differential desiccation between the upper and lower surfaces. Six years later Austroc did realize that gravity is the important factor, but attributed its action to differential effects on the nutrition of the upper and lower sides when the organs were horizontal. (DuHamel disagreed with Austroc's explanation on the basis of the simple observation that a completely inverted shoot still exhibits upward curvature.)

2.2.2   Nineteenth-century investigations

The study of tropisms in general, and of the effects of gravity in particular, was greatly advanced in 1806 by T. A. Knight, an agriculturalist who introduced a more experimental approach rather than relying simply on observations. Knight reasoned that any effects that were due to gravitational force should be capable of being duplicated by the application of centrifugal force. (Strictly speaking, the force acting upon a plant that is being rotated on a centrifuge is centripetal acceleration. However, the term centrifugal force is well established in the literature and also conveys a clear picture of what is happening. Moreover, if the system is considered from the frame of reference of the rotating plant, and with regard to the probable means of stimulus reception, a strong case can be made for using the term centrifugal force, see Hilley 1913.) Accordingly, Knight constructed a wheel that could be rotated at different speeds by the flow of water from his garden stream. Seedlings were attached to the circumference of this water wheel and then subjected to various levels of force. For example, rotation of the 11-inch wheel at 150 r.p.m. subjected the seedlings to a force of 3.5g; this overwhelmed any influence of gravity, and the shoots grew in towards the centre of the wheel while the roots grew outwards. Furthermore, rotation of the wheel in a horizontal plane at 80 r.p.m. (a force of 1.0g) resulted in the roots growing outwards at an angle of 45° rather than completely horizontally. This was a most important finding since it demonstrates that gravity and centrifugal force interact quantitatively, and are thus probably acting through the same system. [Knight is generally credited with the first spinning experiments on plants, mainly because of his reasoning that gravity and centrifugal force act similarly. But John Hunter, died 1793, had previously investigated the effects of rotation on radicle and plumule orientation. He grew a bean seedling on a cylindrical basket which was placed horizontally in notches cut in the rim of a water barrel. A string was wrapped round the plant basket and attached to a float in the water of the barrel. Rotation of the basket was brought about by the water flowing out of the barrel (in Larsen 1962a).]

Throughout the 19th century the study of tropisms progressed rapidly. In 1829, Pinot found that seedling roots could grow down into dense liquid mercury, indicating that roots exert an actual downwards force rather than just depending on their own weight. In 1832, De Candolle gave the name 'heliotropism' to the phenomenon of plant curvature towards light, and, from considerations of the greater elongation of etiolated plants, conjectured that heliotropism was caused by a light inhibition of growth. (The phenomenon was renamed 'phototropism' by Oltmans in 1892.) Hofmeister, in 1863, used the terms 'positive' and 'negative' heliotropism to describe growth towards and away from light, and in 1868 Frank applied the term 'transverse' heliotropism to the horizontal orientations of certain leaves and runners which grow vertically in darkness. Frank also introduced the term geotropism. [In the 1970s, the less euphonius 'gravitropism' became popular. This results in mixed roots of Latin (*gravus* = heavy) and Greek (*tropus* = curved) within the same word. As pointed out by Larsen (1962a), the correct, though perhaps uninformative, term should be 'barytropism', from the Greek *barytes* = heavy.]

The researches of Frank represent another important conceptual milestone, since he considered heliotropism (i.e. phototropism) and geotropism (i.e. gravitropism) to be essentially similar phenomena of differential growth in response to external stimuli. Around the end of the century, tropic responses to other stimuli such as wounding, touch, and chemicals were noted, but the responses to light and gravity received the greatest attention.

In the 1870s two investigators in particular made notable contributions to the understanding of tropisms. The German botanist Sachs, besides his many other contributions to plant physiology, invented the 'klinostat' (see Ch. 3), an instrument that has played an important role in research on geotropism. Further, through detailed growth analyses Sachs demonstrated that in roots the tropic response occurred in the elongation zone behind the root tip and that it was due to a relatively greater growth rate on the convex side of curvature. Meanwhile, around the same period, tropisms and other forms of plant movement were being subjected to the formidable powers of observation and analysis of Charles Darwin. Darwin considered that circumnutation, the spiralling motion of the tip of a growing organ, was the basis of all forms of plant growth movements, and that a tropic stimulus caused this circumnutatory movement to be emphasized in one plane or direction ('natural selection' applied to growth?). Be that as it may, Darwin's most significant contribution to this field was his distinction between reception of a stimulus, and response to the stimulus. In fact, in tropisms he considered these two processes to take place in different regions, with reception of a tropic stimulus occurring in the tip of an organ. In 1872, the Polish investigator Ciesielski had suggested that the root cap was the sensing organ for root geotropism, and by 1880 Darwin, too, was impressed by the sensitivity of the root tip (he drew the analogy, self-admittedly somewhat fancifully, between this region in a plant and the brain in an animal). Thus, Darwin concluded that there must be transmission of some sort of 'influence' between the site of stimulus reception and the region of tropic response, a conclusion that still has a profound effect on present notions of plant growth regulation. (Interestingly, Sachs vigorously disagreed with most of Darwin's conclusions about plant tropisms, including the ideas of tip reception and message transmission. In fact, Sachs considered Darwin to be a bit of a dabbler in plant physiology and a poor experimentalist, although he later accepted Darwin's son Francis into his laboratory in Wurzburg in Germany, to continue investigations on tropic sensitivity in shoot and root tips.)

2.2.3 Mechanisms of stimulus reception

Ideas about the actual mechanisms of reception of tropic stimuli were also being developed by the end of the 19th century. In 1886 Berthold proposed that gravity could bring about a passive sinking of the heavier bodies inside a cell. In 1892 Czapek suggested that starch grains may fulfil this function in plant cells, and in 1900 this idea was independently elaborated by Haberlandt and Němec into the 'starch-statolith hypothesis' to account for gravitropic reception in plants. This asserts that certain cells, statocytes, contain starch grains, statoliths, which move within the cell under the influence of gravity and, through their inter-action with some other cell component, provide the cell with information about its orientation. This concept survives to the present, although there continues to be controversy about its validity and the nature of the statoliths (see Ch. 3).

An indication of the type of molecule that is the receptor for a photobiological process can often be obtained from knowledge about the type of light that best elicits the response (see Ch. 4). In 1817, using a prism, Poggioli showed that the 'more refringent' (shorter wavelength) light rays were the most effective at inducing heliotropism, and this was confirmed in 1843 by Zantedeschi's demonstration that yellow, orange, and red rays were ineffective. In 1878 Weisner constructed a crude action spectrum for phototropism that indicated peak effectiveness in the blue region of the spectrum. (He also reported that a weaker response was induced by the invisible infra-red radiation that had passed through a filter of an opaque solution of iodine in carbon disulphide.) In 1916 A. H. Blaauw produced a more refined action spectrum that confirmed maximum effectiveness of the blue end of the spectrum. Blaauw also introduced a more quantitative approach to the study of phototropism by investigating its (light-)*dose* × (biological-)*response* relationships. He established that the photochemical law of reciprocity held for photo-tropism, in that the same level of response (degrees curvature) was produced by a particular light dose, whether the dose was given as a long exposure to dim light or a short exposure to more intense light. This suggests that only a single type of photoreceptor is involved in the process. However, the actual identity of the receptor is still unknown, although carotenoid or flavin-type molecules remain the obvious candi-dates (see Ch. 4).

2.2.4 *Mechanisms of response*

His studies of the quantitative relationship between light dose and growth response led Blaauw to conclude that light had a direct effect on growth. Thus he suggested that phototropism simply resulted from the differential effects on growth of the different amounts of light received at the opposite sides of a unilaterally irradiated organ. However, research on tropisms at that time was dominated by Darwin's concept of a chemical messenger that acted as an intermediary in the regulation of the growth response. This line of thinking led eventually, through what has come to be viewed as a classic series of experiments by several investi-gators, to the discovery of the plant hormone, auxin, and to an expla-nation of tropic behaviour in terms of a gradient of this growth regulator across the responding organ.

The major organ used in tropic research during this period was the coleoptile of the grass seedling (see Fig. 2.3 and Box 2.1). Around 1910, the Danish botanist Boysen-Jensen studied the tropic responses of coleoptiles which had slivers of mica inserted in such a way as to act as barriers to longitudinal diffusion along one side or the other of the organ; since, for example, there was no phototropic curvature if the barriers were inserted in the shaded side of the coleoptile, the results provided circumstantial evidence that a chemical messenger diffused along the convex side of a curving organ. At the same time, the researches of Paál were emphasizing the importance of the coleoptile tip; removal of the tip prevented any curvature (and growth), and if it was replaced asymmetrically the organ curved away from that side. (The limitations of many of these experiments are discussed later.) In 1923 Söding demonstrated the existence of a growth-promoting substance in plant tissues. Boysen-Jensen considered there was enhanced longitudinal diffusion of this substance along the shaded side of an organ, whereas Paál was convinced that transport was inhibited on the lighted side. At this point, on the basis of his investigations of gravitropism in roots and phototropism in coleoptiles, Cholodny, a Russian, proposed that tropic curvatures resulted from the asymmetric distribution of growth promoter in the tip of a stimulated organ. Around this time, 1920–30, the Dutch physiologist, F. W. Went, was perfecting the technique of 'bioassay' by which amounts of the growth promoter could be quantified. In this technique, the substance was isolated from coleoptiles simply by diffusion from severed tips into blocks of agar (a procedure pioneered by Seubert & Stark); the amount of substance was then quantified by measuring the biological response (say, degrees curvature or amount of growth) when the blocks themselves were placed back onto a coleoptile. Various types of bioassays were developed and, although they all have severe limitations, they did provide a means of testing for the hormonal gradient that is the basis of what has come to be known as the *Cholodny–Went theory* of tropic curvature. (Funnily enough, although their names have become inextricably linked throughout the literature of plant growth physiology, Cholodny and Went never actually met.) The Cholodny–Went theory has several basic tenets:

(a) *Auxin and growth.* The growth of a plant organ is dependent on auxin, which is synthesized in the organ tip.
(b) *Tip perception.* A directional stimulus brings about an asymmetric redistribution of auxin in the tip of the organ.
(c) *Message transmission.* The asymmetry in auxin distribution is transported longitudinally along the organ to produce a response in another region.
(d) *Concentration dependence.* The tropic response represents a growth

    response to altered auxin levels, involving growth stimulation and growth inhibition on opposite sides of the responding organ.

(e) *Differential sensitivity of shoots and roots.* The opposite responses of shoots and roots to gravity are a result of their different sensitivities to auxin.

Subsequent tests of this hypothesis have centred mainly on demonstrations of auxin gradients (or of their absence) across tropically stimulated tissues. These have been carried out by many investigators, most notably in the 1950s and 1960s by K. V. Thimann in the USA, and later by M. B. Wilkins in the UK. Such investigations have progressed from simple bioassay of amounts of naturally occurring auxin, to the demonstration of gradients developed after the exogenous application of radiolabelled synthetic auxins.

    The means by which an auxin gradient could actually be created was also considered in the 1950s. A. W. Galston suggested that auxin may be photodegraded on the lit side of a unilaterally irradiated organ, on the basis of his observation that a solution of riboflavin sensitized the *in vitro* photo-oxidation of indole-3-acetic acid (IAA, a flavin-type molecule being one of the candidates for the role of phototropic receptor). However, quantitative investigations by W. R. Briggs showed that as much total auxin diffused out of irradiated coleoptile tips as out of tips maintained in darkness. Therefore opinion continues to favour some form of lateral migration of auxin in a tropically stimulated organ.

2.2.5 Modern developments

In recent years the Cholodny–Went theory has been seriously challenged by a number of investigators, not only on the basis of criticism of some of the early experiments but also in the light of accumulating results which seem inconsistent with the original theory.

    The early classical experiments have more or less become accepted as dogma over the years, but many of them are flawed, both technically and conceptually. For example:

(a) The early work depended largely on bioassays, but this technique is severely limited by questions of interpretation and specificity. (The technique is rarely used nowadays, but results obtained by it are still often quoted.)

(b) More seriously, most early experiments lack a statistical criticalness, in terms of sample numbers, standard errors, estimates of reproducibility, and statistical significance, etc.

(c) Still more seriously, in early experiments, and often in many modern ones, the effects of wounding are generally discounted or ignored, yet in excision experiments, in barrier insertion experiments, and in work with isolated tissue segments, wounding is a very significant factor. It is well known that a decapitated coleoptile never fully regains its original growth rate, whether the original tip is replaced or exogenous auxin is supplied (Trewavas 1981). Again, the effects of excision and barrier experiments are traditionally interpreted as interfering with the supply of auxin, but other interpretations are possible (Hanson & Trewavas 1982) and in any case, the actual pattern of the changes in growth rate that result from excision of the tip do not support the traditional interpretation (Parsons *et al.* 1988). And finally, wounding itself generally induces directional trauma-tropic growth responses (see Ch. 6), thereby also making the results of the classical barrier experiments open to other interpretations.

Results which seem inconsistent with the Cholodny–Went theory are discussed in more detail in Chapters 3 and 4 (see also Firn & Digby 1980, Pohl & Russc 1984, Firn & Myers 1987). However, in summary they involve three particular aspects:

(a) The presence of the coleoptile tip is not necessary for either gravitro-pic or phototropic responses.
(b) It is claimed that auxin gradients are not created rapidly enough, or are not present, in many instances of tropic curvature.
(c) Analyses of the growth events during curvature have led to suggest-ions that they occur with a rapidity and pattern that is inconsistent with the regulatory involvement of auxin.

The background to each of these criticisms is arguable, and indeed much is still being vigorously contested in the literature. However, although auxin is obviously involved somehow in the tropic responses of higher plants, it is clear that there are experimental inconsistencies and difficulties with the original Cholodny–Went proposals. Another major difficulty may lie in the unification of tropic mechanisms that is attempted by the Cholodny–Went theory. There may be differences between roots and shoots in mechanisms of gravitropism (Ch. 3). And there must be differ-ences between gravitropic and phototropic mechanisms, since it is already clear that light can bring about phototropic curvatures by induc-ing one or more of several different types of growth responses (Ch. 4).

Thus, the regulatory basis of tropisms is in the process of being severely re-evaluated. It may be that, just as the original studies of the directional responses of plants by Darwin and his successors led to the discovery of auxin and its role in the chemical regulation of plant growth, so this re-evaluation may lead to greater understanding of the mechanisms by which plant growth in general is controlled and regulated.

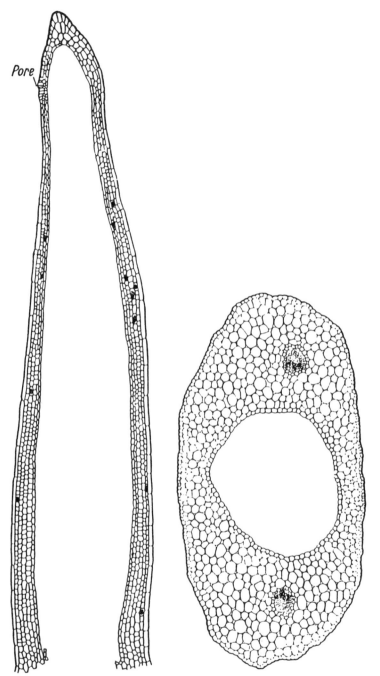

Figure 2.3 Longitudinal and transverse sections of a young oat coleoptile (from Avery &
Burkholder 1936).

Box 2.1   The grass coleoptile

The use of the coleoptile for much of the research on tropisms is in large part a legacy of the work of Charles Darwin, who studied this organ extensively in his general investigation of plant movements (Darwin 1880). Darwin used coleoptiles of canary grass (*Phalaris canariensis*) and, to a lesser extent, oats (*Avena sativa*). Most modern workers have used oats or maize (*Zea mays*), although in the 1920s Paál used the giant coleoptiles of *Coix lachrymajobi* for the technically demanding experiments that involved the asymmetric replacement of excised coleoptile tips.

The coleoptile (Fig. 2.3) is a hollow sheath of mainly parenchymatous cells, with a tiny apical aperture through which the primary leaf bursts when the seedling reaches the surface of the soil. Two vascular bundles run the length of the organ. The coleoptile surfaces are covered with cuticle and, perhaps surprisingly in what is commonly thought of as an underground organ, there are two rows of stomata along the length of the coleoptile in the regions opposite the vascular bundles.

It is worth bearing in mind that the auxin supply of a particular type of coleoptile may originate from quite different tissue sources (Jackson & McWha 1984); in maize most of the auxin in the coleoptile seems to be synthesized in the tip of the organ, but in oats the bulk of the auxin may be supplied from the seed tissues. [It is not clear whether these suspected differences in sources of endogenous auxin are also responsible for differences between the two species in regard to some aspects of their tropic behaviour (Gardner *et al.* 1974); in maize, unilateral light had no effects on the lateral or longitudinal transport of exogenously supplied IAA; but in oats it inhibited longitudinal transport.]

The coleoptile is an ephemeral organ that soon stops growing and dies under natural light conditions, behaviour that at first glance may make it seem an odd choice of material in which to study growth responses, particularly responses towards light. However, the coleoptile functions to protectively sheath the primary shoot axis during upward penetration through the soil towards the aerial environment. It is therefore of considerable adaptive advantage that it possesses a high degree of directional sensitivity in regard to both light and gravity. Other research advantages of the coleoptile are that it is easy to grow and manipulate during experiments, and it is an anatomically simple organ in which cell division ceases early (Avery & Burkholder 1936); after reaching about 1 cm in length, all subsequent growth is by cell extension. [In the equivalent experimental material in dicots, the seedling hypocotyl (Havis 1940), growth is also largely due to elongation.] One practical point is that in oval-sectioned coleoptiles, such as that of oats, the organ curves more readily and more uniformly if it is the narrow side rather than the broad side that receives the phototropic stimulus (Galston 1959).

CHAPTER THREE

Gravitropism

3.1 GENERAL INTRODUCTION

3.1.1 Definition and terminology

Gravitropism comprises those responses in which the direction of growth is determined by exposure to a field of mass acceleration. This type of stimulus is represented in the natural world, of course, by gravity but it can also be provided by the centrifugal forces developed during rapid rotation.

In studies on gravitropism the vertical plumb line is used as the standard reference orientation, and the position naturally assumed by a plant organ or part of an organ in relation to the vertical is called the *liminal angle*. The value of this angle therefore characterizes different types of gravitropic responses. It is simply 0° and 180° respectively in positive and negative orthogravitropism where orientation is exactly along the plumb line, and 90° in diagravitropism, but it is particularly useful for describing the oblique orientations of plagiogravitropic organs, with, for example, 45° describing the orientation of a positively plagiogravitropic root. The liminal angle is characteristic of a genotype rather than of a species, and there can be great variation between different varieties of a plant. It can also be markedly affected by other agencies, such as light, temperature, carbon dioxide, and ethylene. However, the liminal angle of a particular root is generally constant for a particular plant in a particular situation (Rufelt 1969).

Although gravity is continuously and universally present in all natural habitats, its usefulness as a directional signal for plant growth is tempered by the fact that its very generality may not indicate the optimal orientation for growth in a local situation. For example, the action of gravity alone would simply orient organs in a particular direction regardless of whether better conditions of light or nutrition were actually available in another direction. It is worth keeping this major qualification in mind when considering either the general distribution of graviresponsiveness among different types of plants, or the significance of gravitropism in the orientation of an individual plant.

3.1.2 Responses in lower plants and fungi

Gravitropism is found to some degree among all classes of the plant kingdom, although its relative importance as a guidance mechanism does show considerable variation (Banbury 1962). In thallophytes, light is generally the major orienting factor and gravitropism is only of some significance in certain sessile algae. For example, the anchoring rhizoids of *Chara*, *Vaucheria*, and *Caulerpa* are positively gravitropic; additionally in *Chara*, the shoots are negatively gravitropic. In bryophytes, gravitropism is more widely apparent. Liverwort thalli show weak negative responses in the absence of light, and the rhizoids are generally positive; sporophytes of different species vary in their degree of responsiveness. In mosses, however, most shoots and sporophytes are negatively gravitropic. In pteridophytes, light remains the major orienting influence, particularly for the gametophyte generation, but organs of the sporophyte, such as fern fronds, develop fairly strong negative gravitropism with increasing age.

In fungi, vegetative hyphae are generally insensitive to gravity, but many reproductive structures show strong negative responses. For example, sporangiophores of zygomycetes such as *Mucor* and *Phycomyces* are negatively gravitropic (those of *Rhizopus*, however, seem to be insensitive). In ascomycetes, explosive spore dispersal mechanisms are more common, and the general dominance of phototropism in the orientation of reproductive structures in this group suggests that response to light is a more effective means of ensuring that spores are ejected into air currents. Fungal responses to gravity are most apparent, and therefore most extensively investigated, in basidiomycetes. Vegetative hyphae remain insensitive to gravity, but reproductive fructifications show marked gravitropism. The close packing of the spore-bearing gills in a mushroom, for example, means that they must be aligned exactly in the vertical plane in order for the spores to fall free of the cap. This exact alignment is achieved by the negative gravitropism of the stipe (or main stem) acting as a form of 'coarse' adjustment, while the strong positive gravitropism of the individual gills provide any necessary 'fine' adjustment (Banbury 1962).

The motile gametes of algae, ferns, and fungi often show some form of negative gravitaxis (upward swimming) in the absence of any other orienting stimuli (Carlile 1980). Motile responses to gravity have been more extensively investigated in protozoa (Bean 1984). However, there is controversy as to whether gravitactic orientation involves some form of physiological gravity-sensing mechanism within the cell, or whether it is simply a physical effect that arises from the interaction of the gyrational propulsive forces and the gravitational sedimentation forces that are acting upon the organism.

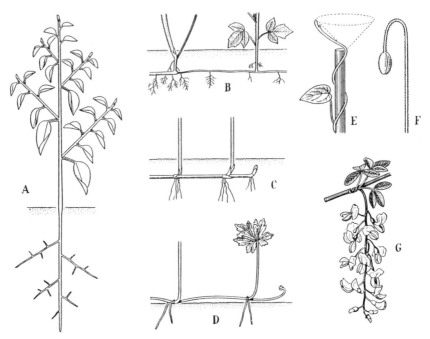

Figure 3.1 Various forms of gravitropic responses in higher plants. (a) Typical dicot, with orthogravitropic main axis (positive root and negative stem), plagiogravitropic branches and secondary roots, and agravitropic tertiary roots; (b)–(d) diagravitropism in roots (*Rubus*), rhizomes (*Heleocharis*), and runners (*Ranunculus*); (e) circumnutation and 'lateral gravitropism' in a twiner (*Ipomoea*); (f) positive gravitropism in a shoot (*Papaver* peduncle); (g) agravitropism in an inflorescence (*Laburnum*). (From Larsen 1962a.)

3.1.3   General responses in higher plants

Gravitropism is a feature of most organs in higher plants, although the form of response and its strength of expression varies markedly according to the type of organ and the conditions. The wide variety of forms of gravitropism in higher plants is illustrated in Figure 3.1. Young rapidly growing seedlings show the most marked responses. Mature tissues are still sensitive to gravity but usually take longer to develop growth responses.

Shoots
Shoots generally show some form of negative gravitropism. [Many young shoots however, particularly coleoptiles, show an initial slight downward bend when first placed horizontal, before developing the normal upward curvature (Pickard 1985a). This early 'wrong way' curvature, as it is termed, is a definite physiological response rather than simply an effect of physical sagging, but its basis is unclear.] As discussed

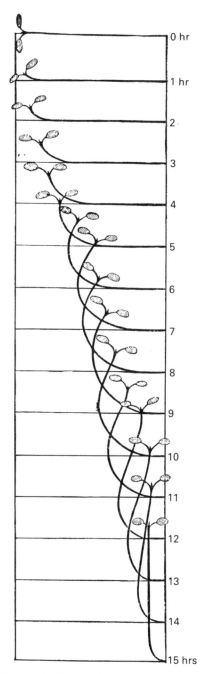

Figure 3.2 Diagrammatic record of the development of gravitropic curvature in a typical dicot shoot (from Pfeffer 1906).

more fully later, gravitropic sensitivity is present along the whole length of a shoot, although the response usually occurs first in the apical region and subsequently migrates towards the base (Fig. 3.2). [Under some conditions, perhaps involving growth of seedlings in total darkness (Pickard 1985a), curvature can occur in a more uniform fashion along the whole length of the responding organ.] The sequence in Figure 3.2 also illustrates some other characteristic features of gravitropic curvature in shoots. Often the apex of the organ overshoots the plumb line. The correction back to the vertical is associated with the autotropic straightening of the organ, or 'counter reaction', which also starts at the tip. The growth responses underlying these changes in orientation are described later.

In the shoots of grasses, the graviresponse is confined to particular regions of the stem. These characteristically swollen regions at the sites of insertion of the leaf bases onto the stem have been variously termed 'nodes', 'pulvini' (because of their involvement in a motor response), and 'false pulvini' (Kaufman *et al.* 1987). In festucoid grasses (e.g. oats, barley) the base of the leaf sheath comprises the swollen tissue; in panicoid grasses (e.g. maize) it is the base of the internode itself that is swollen. In both types, the graviresponse involves the initiation of growth on the lower side of the swollen region, within 15 minutes or so of the organ being displaced from the vertical (Kaufman *et al.* 1987). Each nodal region seems able to act as an independent unit, inasmuch as the reception of the gravity stimulus and the growth response both occur within the same region, but in a complete stem each unit responds to a different extent so that the overall effect is to bring the grass flower or ear back to the vertical. A similar sort of nodal graviresponse is also seen in some dicot shoots, e.g. goosegrass (*Gallium arvense*).

Most secondary aerial axes grow at an oblique angle, and in that sense are therefore plagiotropic. However, this oblique orientation is usually not due to the action of just a single 'plagiotropic mechanism', but represents rather the equilibrium reached as a result of the interactions of various processes (Kaldeway 1962). In the first place, many aerial organs show bilateral rather than radial symmetry, and therefore anatomical differences between the adaxial and abaxial sides can provide some physical basis for the different growth patterns that must underlie an angular orientation. Secondly, some degree of epinastic development (see Section 1.3) is often involved. This is thought to be responsible for the so-called 'basal effect' (Kaldeway 1962), in which the more basal a branch is on a stem, the more horizontal it tends to be. Light, too, usually exerts a very strong influence on the orientation of lateral branches, not only through phototropic effects of blue light and directional effects of red light on growth (see Ch. 4), but also through the effects of light on the expression of gravitropism (see Section 3.1.4). And finally, in some

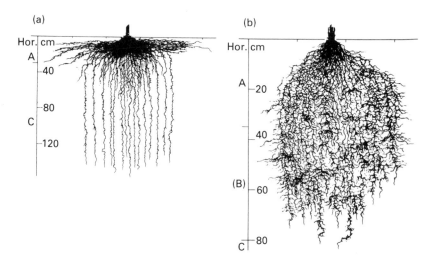

Figure 3.3 Root systems with contrasting morphologies resulting from differences in the proportions of horizontally and vertically growing roots: (a) maize; (b) oats (from Kutschera 1960).

species the formation of 'reaction wood' plays a major role in the orientation of lateral branches (see Section 3.1.5).

Roots

Main roots generally show positive orthogravitropism. The gravity response in a root is localized in the region of cell elongation a few millimetres behind the tip. Thus, when a young primary root is placed horizontally, the tip region grows through 90° to form a permanent right angle. The precise form of the angle varies according to the type of root (Pickard 1985a) and the growth medium (Rufelt 1962). (The differences between roots and shoots in their styles of growth, and therefore in their regions of gravity response, are probably of significance in relation to the different environments of the two types of organ; roots grow in a relatively stable subterranean situation, but must twist and turn to penetrate the soil; shoots, on the other hand, exist in an environment in which circumstances can change drastically, say as a result of a fall, in which case the whole organ must undergo reorientation.)

Secondary roots usually show some kind of plagiogravitropic behaviour, growing out from the main root at their characteristic, but not immutable, liminal angle (Rufelt 1962). Tertiary roots, however, are generally gravitationally neutral, or 'agravitropic'. These different forms of gravitropic response in the different hierarchies of roots combine into a more effective space-filling system for mining the soil of water and nutrients, than if all roots simply showed orthogravitropic responses.

The behaviour of lateral roots is not always clear cut, however. The liminal angle can be influenced by the age of the lateral roots themselves (Moore & Pasieniuk 1984), by their length (Kaldeway 1962) or by their position on the main stem (Rufelt 1962). And in some species, e.g. *Phaseolus vulgaris*, the lateral roots seem to be completely unresponsive to gravity (Ransom & Moore 1983). Such differences in the responses of roots towards gravity can result in quite contrasting morphologies of root systems (Fig. 3.3). However, as has been cogently pointed out (Jackson & Barlow 1981), research in this area has concentrated upon the mechanisms responsible for the actual process of root realignment (and simply realignment to the vertical position at that), while largely ignoring the means by which the liminal angles are maintained and by which the characteristic 'gravitropic profile' of the root system is developed.

More specific aspects of plagiotropism and diatropism in shoots and roots are discussed in the following section.

3.1.4 *Variations in response*

In higher plants there is considerable variation in gravitational responses, and a particular type of organ by no means always shows the standard type of response. Potato stolons, for example, are stems that are positively gravitropic; aerial branches of 'weeping' tree forms, such as those of willow (*Salix* spp.), have a slightly positive response to gravity; and the 'pneumatophores' of mangroves are negatively gravitropic roots which grow upwards out of the water to act as air-breathing snorkels. Again, the response to gravity can differ in different regions of an individual organ. This is seen in the behaviour of the plumular hook in certain dicot seedlings. This hook is not preformed in the embryo within the seed but develops during the early growth of the seedling and, at least in cress (*Lepidium*), sunflower (*Helianthus*), and cucumber (*Cucumis*), its formation is dependent on gravity. That is, although the major part of the hypocotyl is negatively gravitropic, the apical region is positively gravitropic and grows over into a downward-pointing arch (MacDonald *et al.* 1983a). And finally, some organs are wholly agravitropic, such as the hanging inflorescences of *Laburnum* (Fig. 3.1g). (Many of the so-called 'lazy' mutants, e.g. of rice, maize, and tomato, in which the mature plant grows prostrate along the ground, are really agravitropic forms that simply grow in whatever directions their organs happen to be pointing.)

Variations in the response to gravity can also be brought about by the actions of other factors. These factors can be of an endogenous developmental nature, or they can be environmental.

Developmentally regulated variations in gravitropism

One of the most extreme and dramatic examples of a developmentally regulated change in the gravitropic response of an organ is seen in the behaviour of the gynophore (or ovary-bearing stalk) in the flower of peanut (*Arachis hypogea*). After fertilization, the gynophore resumes growth and becomes positively gravitropic to the extent that the ripening seed head is carried downwards and buried in the soil (Jacobs 1947).

Less extreme examples of developmentally regulated changes in the gravitropic responses of an organ can be found in some of the flower species which exhibit 'nodding' behaviour, i.e. the flowers undergo marked changes of upward or downward orientation at different stages of development. Many of these changes in orientation are best considered as 'gravi-epinastic' effects (Kaldeway 1962, 1971), in which the onset of the differential growth responsible for bending is induced by gravity but the actual pattern of growth and mode of bending is determined by factors within the tissues. In poppy (*Papaver rheo*), the straightening of the stalk after the downward orientation of the flower bud stage is due to the gradual onset of a strongly negative gravitropism in the apical regions of the stalk (Kohji *et al.* 1981). (The subsequent drooping of the seed head stage, previously described as positive gravitropism, has now been reported by the same authors to be simply a weight effect of the seed pod.)

Examples of endogenously regulated changes in organ orientation that involve effects of gravity, though not necessarily in a strict gravitropic sense, are seen in the behaviour of lateral branches in conifers. In *Pinus* spp., the horizontal orientation of a lateral branch (its 'equilibrium position'), results from the interaction between a downward-acting influence from the main apical bud (analogous to an epinastic effect), and an upward-acting gravitational effect on the lateral branch itself; the gravitational effect operates through the induction of 'reaction wood' on the underside of the branch (see Section 3.1.5). Removal of the leader shoot of a tree thus results in the removal of the apical effect, and the lateral branches take up a more vertical orientation until one of them assumes the role of leader shoot (Wilson 1973). This response of the lateral branches to removal of the apical bud has been variously described as the development of negative gravitropism in lateral branches (Audus 1969), hyponasty in lateral branches (Palmer 1985), and a gravitational effect on the formation of reaction wood (Fisher 1985).

Potato stolons, too, tend towards an upward direction of growth if the apical bud of the main stem is damaged or removed. However, again it is not entirely clear whether this represents the development of a true negative gravitropism or simply the onset of hyponastic growth.

Environmentally regulated variations in gravitropism

Light is the primary environmental factor that exerts the greatest and most general effects on gravitropism. Not only do the phototropic effects of light generally override gravitropic responses in the natural environment (see Section 4.4), but also brief treatment with light usually changes the nature of the gravitropic response itself.

In roots in general, light increases gravitropic sensitivity (Feldman 1984, Pickard 1985a, Woitzik & Mohr 1988b), and in certain varieties of maize it transforms the unusual plagiotropism of the primary roots to strong positive orthogravitropism (Wilkins 1979, Mandoli *et al.* 1984). The adaptive significance of such responses is fairly obvious but the underlying mechanism is far from clear. Light promotes the synthesis of root growth inhibitors, including abscisic acid (ABA) (Feldman 1984), and at one time this was thought to be the means through which it affects gravitropic sensitivity (i.e. by enhancing the level of the inhibitor that was responsible for regulating the differential growth of gravitropism). However, it is now fairly well established that ABA is not the primary regulator of gravitropic growth (Jackson & Barlow 1981, Moore & Evans 1986). Other suggestions to account for the light enhancement of gravitropic sensitivity include possible effects on other inhibitors (Suzuki *et al.* 1979), or on the interaction of auxin and ethylene (Feldman 1984), or via secondary effects of ABA on the gravitropic system (Pickard 1985a).

Until recently the effects of light on gravitropic responses in shoots have been more contentious, with some studies reporting light enhancement of sensitivity to gravity and others, a light-induced decrease in sensitivity. However, it has now been shown that the effects of red light are related to the direction of irradiation (Woitzik & Mohr 1988b). If red light and gravity are acting in the same direction, gravitropism is enhanced; in other circumstances, red light decreases gravitropic sensitivity. This means that in shade conditions (i.e. low amounts of red light) gravitropic responses are enhanced (Woitzik & Mohr 1988b). Again the adaptive significance of such a system is clear, but, except that phytochrome is obviously involved, little is known of the underlying mechanisms.

Light overrides the positive gravitropism that is responsible for the formation of the hypocotyl hook in many seedlings (MacDonald *et al.* 1983a), and thus brings about hook opening when the seedling breaks through the surface of the soil. This, too, is a phytochrome-mediated response.

Light also affects the plagiogravitropic behaviour of many horizontally growing organs (described later). Some effects in this areas, however, are indirect; in couch grass (*Agropyron repens*) exposure of the parent plant to high-intensity light maintains diagravitropism in the rhizomes, while under low-intensity light the rhizomes grow upwards (Palmer 1958).

More directly, high light intensities induce positive plagiogravitropic responses in the shoots of many stoloniferous grasses (Palmer 1956). This is suggested to be the general explanation for the transformation of the erect growth habit of many plants in shade conditions, to a prostrate form in sunlight [and disputes an earlier claim (Langham 1941) that the prostrate habit was due to negative phototropism at high light intensities].

Temperature, too, can affect the gravitropic behaviour of some species. For example, at low temperatures stems of *Lamium purpureum*, *Sinapis arvensis*, and *Senecio vulgaris* tend to become plagiogravitropic (Coombe & Bell 1965). The resulting prostrate growth habits are presumably of adaptive advantage in helping these common weed species to survive adverse winter conditions. Again, the underlying mechanisms are not clear.

Mechanical stimulation is another important factor that can modify gravitropic responses. In fact, in roots either mechanical contact (Darwin 1880) or friction (Edwards & Pickard 1987) can each completely overrule the gravitropic response. This behaviour is obviously important in enabling roots to grow around obstacles and to penetrate the line of least resistance through the soil. These thigmotropic responses are discussed in Chapter 5.

Thus, gravitropic responses are capable of being greatly influenced and changed by many other factors in the environment. This situation gives substance to the qualifying remarks which introduced the chapter. That is, gravity is constantly present as a directional signal, but it provides information of such a general nature that its effects must often be modified according to local circumstances.

Ethylene and organ orientation

Ethylene has very great effects on the orientation of plant organs. Therefore, since increased ethylene synthesis is a common response to situations or agencies that impose some kind of stress on the plant, such situations often have marked effects on organ orientation. Examples include changes of orientation in response to mechanical stimulation, physical disturbance, reorientation, exposure to high light intensities, and the onset of senescence (Palmer 1985, Pickard 1985a).

There are several possible means through which ethylene could influence the orientation of an organ:

(a) It has been suggested that ethylene may play a direct role in gravitropism by acting as a primary regulator of differential growth in some systems (Wheeler & Salisbury 1981); this aspect is discussed further in Section 3.3.2.

(b) Ethylene may interact in gravitropism to slow down, or even completely prevent the normal responses of an organ (Pickard 1985a).

(c) Ethylene may induce a nastic response (Palmer 1985). In this case the form of growth, and therefore direction of orientational change, is specific to the organ. It can involve epinastic (downward) growth, e.g. lateral branches, most diagravitropic rhizomes, and some flower stalks; or hyponastic (upward) growth, e.g. some diagravitropic runners (strawberry, clover) and the prostrate tomato mutant 'diageotropica'.

Overall, this means that a very wide range of agencies, all acting through increased ethylene synthesis, can produce a bewildering variety of changes in orientation among different organs.

The mechanism by which ethylene brings about these effects is not entirely clear. Ethylene and auxin are generally considered to be balancing members of a feedback system, in which auxin stimulates ethylene production and ethylene inhibits auxin synthesis and transport (Palmer 1985, Pickard 1985a). Therefore any change in circumstances that affects any aspect of the system, say auxin transport or ethylene synthesis, upsets the equilibrium and induces growth responses that are specific to the particular species or organ.

Plagiotropism and diatropism

It has already been pointed out that the plagiotropic orientations of aerial organs result from the interactions of several processes. In the 1940s, Snow considered that the angles of lateral branches represented the equilibrium between an influence from the apical bud, which tended to force the branch down towards a horizontal position, and the innate negative gravitropism of the branch itself. More recently, the phenomenon of epinasty has taken the place of the apical influence as the downward-acting tendency (Kaldeway 1962, Palmer 1985). This view derives from the finding that where plants are grown for long periods on a rotating klinostat, which tends to minimize unilateral gravity stimulation (see Box 3.1), the lateral branches tend to assume a downward curving orientation. (This, however, could be due to an enhanced production of ethylene as a result of the klinostat treatment.) In any event, in the case of aerial organs there seems to be no clear distinction between 'plagiogravitropism' and 'gravi-epinastic effects' (Pickard 1985a) in that some kind of endogenously regulated form of differential growth acts to force a branch downwards, while a gravitropically induced pattern of growth tends to force the branch upwards. The particular balance reached is presumed to be specific for a particular organ under particular conditions.

The situation may be somewhat different in underground organs. In

54

(a)

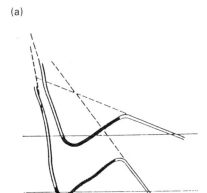

(b)

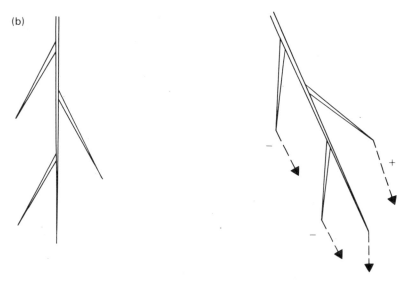

Figure 3.4 Diagrammatic representations of plagiogravitropic responses in secondary roots: (a) secondary roots of potato grown first in the normal orientation, then inverted, and finally returned to the original orientation (from Sachs 1887); (b) positive (+) and negative (−) gravitropic responses in secondary roots of a plant that has been placed at an oblique orientation.

secondary roots, plagiotropism can be demonstrated in two ways. Firstly (Fig. 3.4a), if a plant is completely inverted the secondary roots change their direction of growth to regain their liminal angle, and this can occur through several cycles of inversion and reinversion (Rufelt 1962).

Secondly, and perhaps more surprisingly, if a main root is placed at an oblique orientation (Fig. 3.4b) so that the secondary roots on the lower side are at angles less than their liminal angle, and the secondary roots on the upper side are at angles greater than the liminal, then the former secondaries show negative gravitropism and the latter show positive gravitropism in order to recover the liminal position (Rufelt 1962). This ability of secondary roots to show either negative or positive gravitropism depending on how they are displaced from their liminal position was first demonstrated by Czapek in 1895 (Rufelt 1969), and is considered to be the true test of plagiotropic behaviour, rather than simply growth at an oblique angle.

The mechanism that is responsible for true plagiogravitropic behaviour in secondary roots is not known. Subterranean organs generally have a high degree of radial symmetry and therefore there is no anatomical basis for asymmetry in growth pattern. Early ideas simply envisaged plagio-gravitropism as resulting from opposing positive and negative gravitropic tendencies, but there are several difficulties with this. For example, how would this account for the characteristic liminal angle being maintained over considerable periods of time? And, more significantly, how would it account for the ability of a particular secondary root to adjust either positively or negatively to recover the liminal direction? (For example, if it was simply a balance between opposing positive and negative tendencies, each plagiotropic root would have two possible liminal positions, oriented either upwards or downwards.) It was therefore proposed some time ago that the length of the secondary root and its position on the main root also exert regulatory effects, and that the liminal angle is determined by the interactions between positive and negative gravitropic tendencies and another feature termed 'longitudinal force' (Rufelt 1962, 1969). More recently it has been suggested that the plagiotropic responses of lateral roots are the result of their small size which prevents the establishment of an effective concentration gradient of growth regulator (Moore & Pasieneuk 1984).

Diagravitropism is often considered as a special case of plagiotropism. Again, it is thought to result from the opposition of positive and negative tendencies within the same organ (Rufelt 1962), although in many cases there are often effects of other factors. For example, the rhizomes of couch grass (*Agropyron repens*) and the stolons of Bermuda grass (*Cynodon dactylon*) are diatropic only as long as they are attached to their parent plants (Palmer 1958, Montaldi 1969). (This so-called 'induced plagiotropism' in rhizomes seems an analogous situation to the plagiotropism of conifer lateral branches that is induced by the leader shoot.) Again, the horizontal growth of some organs is directly affected by the actions of external factors. For example (Audus 1969), rhizomes of ground elder (*Aegopodium podagraria*) are diatropic in darkness, positively plagiotropic if exposed to light (i.e. grow downwards), and negatively plagiotropic in high concentrations of carbon dioxide (or ethylene?). Thus, in this case the interaction of diatropism with factors that convey information about aerial or subterranean conditions keeps the organ growing horizontally at a fairly constant soil depth.

3.1.5   Other effects of gravity

Effects on turgor movements

Gravity can stimulate turgor-based changes in leaf orientation in some legumes. In *Mimosa*, gravity-induced turgor adjustments within the primary pulvinus can maintain the leaf in a particular orientation (Roblin & Fleurat-Lessard 1987). In *Phaseolus vulgaris* gravity-induced changes in cell turgor also influence leaf orientation, but the nature of the change varies according to the state of the plant's circadian rhythm (Hosokawa & Kiyosawa 1983); if the plant is inverted during the day, the leaves are adjusted in an upwards direction, but inversion at night brings about downward leaf adjustment.

The turgor changes in the pulvini are associated with ion movements. During the gravity-stimulated leaf movements in *Mimosa*, potassium and chloride move to the adaxial half of the pulvinus, and calcium increases in the abaxial half (Roblin & Fleurat-Lessard 1987). Experiments with ion-specific transport inhibitors and chelators suggest that potassium transport is essential for seismonastic leaf movements but not for the gravity-stimulated movements, whereas calcium is involved in both the seismonastic and gravitational movements.

Geostrophism

The term geostrophism is used to describe gravity-induced orientation of an organ by twisting, rather than by (tropic) bending (Snow 1962). Leaf petioles of *Robinia* and *Wisteria*, for example, can twist along their long axis so that the leaf blade is horizontally positioned for maximum light interception; light itself, of course, can also induce such behaviour (photostrophism). Again, in *Philadelphus*, where pairs of leaves are arranged opposite each other, the internodes can twist strophically to minimize leaf overlap. Such strophic movements are based on growth, but are distinguished from tropisms simply because the organ remains straight during its reorientation rather than becoming curved. However, it is not at all clear exactly where or how the asymmetric growth occurs in strophic movements.

Gravimorphism

The general term gravimorphism (geomorphism) is used to describe effects of gravity on aspects of plant form other than orientation. These include such diverse effects as the development of lateral branches only from the upper side of a horizontal branch (Audus 1969), and the induction of flowering in some species in response to a change to a horizontal orientation (Jankiewicz 1971).

The development of the 'cucurbit peg' (Darwin 1880) is another example of a gravimorphic effect. Early in germination most species of the Cucurbitaceae develop a protuberance, or peg, of cortical tissue on whatever is the lowermost side of the seedling axis in the region of the root–shoot transition zone. This peg inserts itself into the seed coat and acts as a lever against which the cotyledons can free themselves from the coat (MacDonald *et al.* 1983a). Similar structures are present in the seedlings of some other angiosperms and gymnosperms (see Witzum & Gersani 1975).

In many species flower shape is influenced by gravity, through the induction of differential growth in different parts of the flower head (Jankiewicz 1971). For example in *Gladiolus* spp. the lower leaves of the perigonium are much larger than those on the upper side, and in *Epilobium* spp. most of the petals are grouped on the upper side of the receptacle.

In *Agropyron repens* leaf shape is influenced by gravity (Palmer 1958). If the rhizome is placed with its tip upwards, the rhizome scale leaves are transformed into regular shoot leaves over a period of 10 days or so.

Formation of reaction wood

The formation of 'reaction wood' is a major factor controlling the orientation of lateral branches in many woody species. It involves the increased production of xylem tissue, and there are two types of response (Wilson & Archer 1977). In conifers, xylem production is enhanced on the lower side of a branch to form so-called 'compression wood' that acts to push the branch up towards a more vertical orientation. In woody angiosperms, on the other hand, more xylem is produced on the upper side and this 'tension wood' tends to pull the branch up towards the vertical.

The nature of the stimulus responsible for the formation of reaction wood is the subject of long and continuing debate (Wilson & Archer 1977). Older views favoured mechanical stresses as the inducing factors. In the 1970s, gravitational effects received more serious consideration in the induction of compression wood (Wilson 1973). Recent studies indicate that in the formation of tension wood, either gravity or tensile stress can act as inducing factors, but that gravitational effects are dominant if the directions of the two forms of stimuli are in opposition (Fisher 1985).

Box 3.1   The centrifuge and the klinostat in gravitropic research

A prime requirement in the physiological investigation of any factor is the facility for controlled application of the factor, particularly its intensity and its duration. Gravity, however, is ubiquitously present over the surface of the Earth and, unlike light for example, it cannot be simply excluded or switched off. Its manipulation for the investigation of its biological effects therefore requires special approaches.

The intensity of the gravitational stimulus can be altered in two ways:

(a)   Cross-sectional intensities of mass acceleration of less than $1g$ can be obtained simply by placing the organ at different angles. In this case, the gravitropic response often follows what is termed the 'sine rule', i.e. response $= 1g \cdot$ sine $\alpha°$, where $\alpha°$ is the angular deviation from the vertical downwards orientation. Thus, the greatest effect is obtained in the horizontal position, since sine $90° = 1$. (However the gravitropic response does not always obey the sine rule, particularly over long periods of stimulation, see Larsen 1962b.)

(b)   Intensities of mass acceleration greater than $1g$ can be obtained by rotating the material on a centrifuge. In this case, $g = (11.2 \times 10^{-6})$ $\times$ r.p.m.$^2 \times$ radius cm (Chance & Smith 1946). Centrifugation studies have been used for several different purposes. Different regions of an organ can be exposed to different intensities of stimulation by centrifuging the organ at different angles and positions on the centrifuge (see Larsen 1962a for details). This technique was pioneered by Piccard at the beginning of the century, in order to determine the regions of gravitropic sensitivity within different organs. Again, the centrifuge has been used in studies of cell polarity and of the possible interactions between gravity-sensing statoliths and other cell components (Volkmann & Sievers 1979). And gravitropism in the *Phycomyces* sporangiophore has been investigated by centrifugation (Dennison 1971); in this last case new insight was gained into a possible mechanism for gravity (centrifugal) stimulus reception that involves mechanical deformation of the cell wall.

The duration of a unilateral gravity stimulus can be controlled by transferring the plant on to a klinostat. Traditionally, a klinostat was a device which rotated a plant horizontally about its long axis, so exposing it to omnilateral rather than unilateral gravitational influence, but some modern machines, so-called orthogonal klinostats, rotate the plant simultaneously in more than one plane (Shen-Miller *et al.* 1968). Such devices have been used to investigate the effects of hypogravity on circumnutation (Brown & Chapman 1988), to determine the threshold values for gravitropic stimulation (Shen-Miller *et al.* 1968), and to investigate gravitational dose–response relationships at intensities of less than $1g$ (Brown & Chapman 1981); in this last case growth effects in different types of plant showed remarkable similarity, with an exponentially declining response up to saturation at around 1g.

(However, klinostat rotation has itself been found to bring about many physiological effects in plants, including loss of cell polarity, changes in growth rate and in tropic sensitivity (Volkmann & Sievers 1979), and, most significantly, increased rates of ethylene synthesis (Palmer 1985). And, of course, ethylene itself brings about many different types of physiological effects.)

Another means of controlling the pervasive gravitational influence is in the microgravity conditions of spaceflight and Earth orbit, and several biosatellite missions have been carried out (Halstead & Dutcher 1987). In plants grown under these conditions, there are usually many signs of physiological damage, including chromosome abnormalities, many ultra-structural changes, and often death at or close to the flowering stage. However, at least one species, *Arabidopsis*, has been grown from germination through to seed production, suggesting that the problems are technical rather than physiological. The *Arabidopsis* success also suggests that although gravity may be an important source of directional information for the plant, its presence is not really crucial to any aspect of plant development. (One unexpected finding was the lack of regeneration of the root cap in Space, suggesting that the formation of the traditional gravity-sensing tissue of the root is itself a response to gravity.)

3.2 STIMULUS RECEPTION AND TRANSFORMATION

3.2.1 *Parameters of stimulation*

In research on gravitropism, certain terms and parameters are used in order to formalize and quantify the relationship between the gravitational stimulus and the biological response.

As with the application of any form of stimulus, the total gravity stimulus, i.e. the *stimulus quantity* or 'dose', is the product of the duration over which the stimulus is applied and the intensity of the stimulus that is provided:

$$\text{dose} = \text{time} \times \text{intensity}.$$

Each of these components of the gravity stimulus is discussed below. (The means by which the duration and intensity of the gravity stimulus can be controlled is discussed in Box 3.1.)

One further point concerns the relationship between time and intensity of stimulation. In a straightforward situation, there should be complete equivalence or interchangeability between these two factors; that is, for a constant dose the same level of biological response should be obtained, whether this dose is due to a low intensity acting for a long time or a higher intensity acting for a short time. Such a situation is said to show 'reciprocity', and the dose is described as the 'reciprocity constant'.

Threshold time

The minimum period of exposure to a gravity stimulus of $1g$ at $90°$ that is required to produce a gravitropic response is called the *presentation time*. Its value gives an indication of the boundary limits within which events involved in stimulus reception must occur, and these are usually of the order of seconds or, at most, a few minutes.

The actual values of presentation times vary according to a number of different factors. They vary between different species, different types of organ, and different stages of development. They also vary according to environmental conditions, particularly temperature, which should be between $20°C$ and $30°C$. And, not least, they vary according to the method that is used to measure them (Audus 1969). Originally, presentation times were determined as the minimum time taken to produce a measurable gravitropic response in 50% of a population, i.e. it was the measurement of the average individual. This '50% method' gave presentation times for cress roots, for example, of 3–5 minutes, depending on temperature. More recently, methods have involved plotting the extent of gravitropic curvature in a population against increasing periods of stimulation, and then deriving the presentation time by extrapolating the graph back to zero curvature, i.e. it is a measurement of the most sensitive individual. This 'extrapolation method' gives presentation times of around 12 seconds for cress roots, 18 seconds for bean roots, and around 30 seconds for oat coleoptiles (Volkmann & Sievers 1979).

Even these short extrapolated presentation times, however, do not completely indicate the great sensitivity and rapidity with which a gravity stimulus is detected. Responses can be obtained by the summation of bursts of stimulation, each one of which is much shorter than the presentation time if the time between each stimulation is not too long, e.g. stimulation for 0.5 second at one-second intervals can induce curvature in oat coleoptiles (Pickard 1985c). Results such as these suggest that tissues can *detect* a change in orientation virtually immediately, although the response obviously takes much longer to become apparent.

Threshold intensity

Threshold angles of orientation which are capable of causing gravitropic curvature are between $0.5°$ to $10°$ from the vertical for shoots of pea, bean, and sunflower (Volkmann & Sievers 1979). As would be predicted by the sine rule (see Box 3.1), the extent of response generally increases with increasing angle from the vertical up to the optimum angle of $90°$ (i.e. horizontal). However, an important qualification is that this relationship only holds for short exposures to gravitational stimulation (Larsen 1962a, b). In longer exposures, say of hours, angles of greater than 90° produce larger responses than 'sine equivalent' angles of less than 90° (e.g. in primary roots, orientation at 135° from the vertical downwards position generally produces a greater response than orientation at 45°, even though sine 135° = sine 45°) That is, during long exposures to stimulation the extent of response seems to become influenced by other factors within the tissues themselves, rather than simply being a function of the stimulus intensity alone.

Measurements have also been made of the minimum acceleration that can induce a response when the stimulus is applied over considerable periods. This has been carried out on an orthogonal klinostat (see Box 3.1) and the threshold intensities for oat coleoptiles and roots were found to be 10^{-3} g and 10^{-4} g, respectively (Shen-Miller et al. 1968). The generally greater sensitivity of roots has been confirmed in studies carried out in the microgravity environment of a biosatellite, which showed lettuce hypocotyls and roots to respond to threshold intensities of 2.9×10^{-3} g and 1.5×10^{-4} g (Halstead & Dutcher 1987).

Threshold dose

Reciprocity between time and intensity has been found to apply for the threshold stimulus quantity that evokes a response in oat coleoptiles, over the wide range of stimulus intensities of 0.09g–12.5g (Volkmann & Sievers 1979). Values for threshold stimulus quantities ('time × intensity') for oat coleoptiles have been obtained by extrapolation from plots of curvature against stimulus, and have been found to vary from 240g · s (at 22.5 °C) to 120g · s (at 27 °C). There is an obvious discrepancy between these estimates of stimulus quantity and the (extrapolated) presentation times of 30 seconds (i.e. a stimulus quantity of 30g · s), which is suggested to be due to the biphasic nature of the extrapolation curve used to estimate stimulus quantity (see Volkmann & Sievers 1979).

Overall, investigation of these parameters of gravitational stimulation has shown that plant organs are extremely sensitive to forces of mass acceleration and can detect very small changes in orientation, if not instantaneously, then certainly within seconds of the onset of stimulation.

3.2.2 Mechanisms of stimulus reception

Regions of sensitivity

It is often stated, on the basis of excision experiments, that the tip of an organ is (or is not) the region of gravity stimulus reception. This is particularly true in the case of roots where, in some varieties of maize, the root cap can be dissected off without damaging the rest of the root tissues

(Juniper 1976). These decapped roots continue to grow but show no gravitropic response for the 24 hours or so it takes to regenerate a new cap. However, such experiments alone simply indicate that the cap is necessary for the response to gravity, not that it is the localized region of sensitivity to gravity (Jackson & Barlow 1981).

Additional evidence that the root apex is the site of gravitropic stimulus reception comes from a variety of other, and earlier, experimental approaches (references cited in Jackson & Barlow 1981):

(a) In 1894 Pfeffer used specially shaped glass tubes to demonstrate that it is the orientation of the tip, rather than the rest of the root, that is important in gravitropism; he allowed roots to grow into pre-bent tubes (so-called Czapek 'boots'), then exposed different regions to different orientations and noted the growth responses after the tubes had been broken open.

(b) In 1902, the experiments of Francis Darwin confirmed the importance of root tip orientation; when the tip was held fixed in a horizontal position, differential growth continued until the root formed a complete loop.

(c) In 1908, Haberlandt used Piccard's technique of differential centrifugation to show that the apical 2 mm of the root is the most sensitive to stimulation; in this technique, different regions of an organ are exposed to different intensities of centrifugal stimulation by being fixed in different orientations on the centrifuge.

(d) Several investigators have since confirmed the importance of tip orientation by noting the responses when excised root tips are replaced at various angles.

The general presence of starch grains that move in response to gravity in certain cells of the root cap also provides correlative evidence that the putative gravity-sensing mechanism, statoliths, is present in the region of greatest gravitropic sensitivity (Moore & Evans 1986). The statenchyma, the statocyte-containing tissue, is considered to be in the columella or central core of the cap (Fig. 3.5a). The cells of this tissue are non-vacuolated and their walls have a higher number of plasmodesmata (Volkmann & Sievers 1979). Until recently cytoplasmic streaming was thought not to occur in these cells, but saltatory movement of the smaller cell particles has been observed in maize columella cells (Sack *et al.* 1986). In some species there is a strong polarity or stratification of ultrastructure in the cells of the columella (Volkmann & Sievers 1979); endoplasmic reticulum (ER) is located at the 'bottom' (root tip) end of the cells, with starch grains layered on top of it. However, this arrangement is not found in root statocytes of all species (Juniper 1976, Moore & Evans 1986).

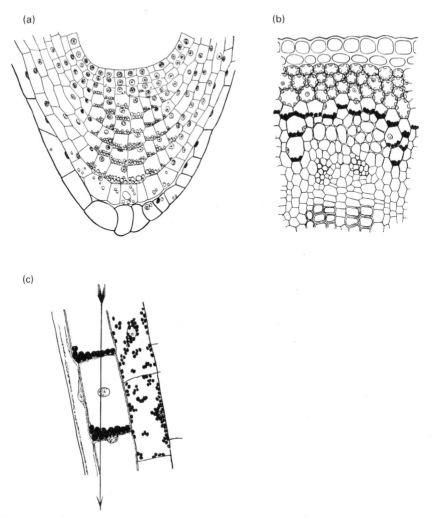

Figure 3.5 Sections through various tissues, showing the distributions of starch grains under different orientations: (a) longitudinal section through a vertical root cap of *Nasturtium*; (b) transverse section through a horizontal internode of *Linum*; (c) radial section through a node of *Tradescantia* after displacement from the arrowed vertical orientation (from Haberlandt 1914).

The regions of aerial organs that show greatest sensitivity to gravity are much more difficult to distinguish. In 1912, Guttenberg used Piccard's technique of differential centrifugation to demonstrate that the apical 4 mm of the oat coleoptile was most sensitive, although a certain degree of sensitivity was present along the whole organ (Larsen 1962a). Furthermore most, if not all, shoots show at least some gravitropic curvature in the absence of their apex. (However, although there is this 'localized' aspect to gravitropism in shoots, the apex and its plane of orientation are important to some other features of the overall response, see Section 3.3.2.)

It is not entirely clear which cells act as statocytes in shoots. The original concepts of Haberlandt and Němec in this area were derived from their clear observations that movement of starch grains in response to gravity occurred only in certain cells or tissues (Fig. 3.5b). Since then it has often been confirmed that gravity-sensitive starch grains are specifically located in the endodermal sheaths around vascular tissues (Parker 1979, Clifford & Barclay 1980, Kaufman *et al.* 1987) or in the cells of the inner cortex (Sliwinski & Salisbury 1984). Such cells typically contain ten or more starch grains, in comparison to the single large grains that are usually present in cells of starch storage organs (Heathcote 1981). However, not all shoots have an endodermis or starch sheath. (There is also an element of circularity in this aspect, since the evidence that starch grains act as gravity-sensing statoliths in the shoot is itself largely circumstantial.) In other respects, these putative shoot statocytes are quite different from those of the root cap, in that they generally have a large central vacuole surrounded by a peripheral layer of cytoplasm that has only a sparse content of ER (Parker 1979).

Stimulus reception

An important point to bear in mind about gravity is that, unlike say light, it does not itself act as a unilateral stimulus. That is, there is no greater intensity of gravitational force impinging on one side of the organ than on the other. Therefore a directional response to gravity must occur through the creation of some kind of physiological asymmetry by the cells or tissues themselves.

Various means have been suggested by which gravity could bring about such an asymmetry in an organ (reviewed in Audus 1969, 1979):

(a) In 1978, Kessler (cited in Audus 1979) suggested that differential cyclosis, in which cytoplasmic streaming was slower in the 'up' direction, would result in an overall downward polarity of materials within the tissues. However Audus (1979) considered that such an effect would be insignificant in relation to the small size of the cells and the large effects of diffusion.

(b) Several investigators (Pickard & Thimann 1966, Hertel 1971, Pollard 1971) have suggested that the weight of the cytoplasm or protoplast itself may bring about differences in the properties of the plasmalemma on the upper and lower sides of a cell, and thus create a physiological gradient across the organ. Again Audus (1969, 1979) considered that not only would such a pressure differential across a cell be insignificant, but also that the small threshold intensities of stimulation that are effective in inducing a response could only, at most, be bringing about minor changes in membrane properties. (However there may be some means of 'focusing' or magnifying the effects of mass acceleration on the membrane, see Edwards & Pickard 1987 and p. 160.)

(c) Detection of gravity by differences in the distributions of cellular components was suggested by Berthold in 1886, a notion that led to the proposal of the starch statolith hypothesis by Haberlandt and Němec in 1900 (see Section 2.2).

However, whether starch grains do in fact function as statoliths in plant cells has been, and continues to be, a fairly contentious issue. On the basis of Stokes' Law governing sedimentation, and by making assumptions about cytoplasmic viscosity, Audus (1969) originally calculated that only a particle at least 0.1 μm in diameter and denser than a mitochondrion, would move across half a cell width (10 μm) within a presentation time of 3 minutes. (If a smaller distance is assumed to be moved, then of course a statolith could be smaller than this; but on the other hand, shorter presentation times would require larger particles.) Actual measurements of the rates of movement of the starch grains after organ reorientation indicate that they can move far enough and fast enough to act as gravity-sensing statoliths, even within the shorter (extrapolated) presentation times (Heathcote 1981). Other evidence that starch grains function as statoliths is almost wholly correlative and covers the following areas (see Audus 1979, Wilkins 1984, Moore & Evans 1986, Kaufman et al. 1987):

(a) There is a correspondence between the presence of starch grains and gravitropic sensitivity in a wide variety of species; this correspondence is particularly strong in the case of the root caps, and even in onion (Allium cepa) and Asparagus, species which do not form storage starch, the cells of the root cap contain starch grains.

(b) There is a close correlation between the effects of temperature on the sedimentation time of starch grains in cells, and the gravitropic presentation time; for example in stems of Lathyrus sp., the fastest sedimentation and the shortest presentation time both occurred at 30 °C.

(c) Starch can be completely degraded by appropriate hormonal treatments, after which no gravitropic sensitivity is apparent until starch reforms; earlier discrepancies in experiments of this type seem to have been due to difficulties in removing 'statocyte-starch' as well as the more easily degradable 'stored-starch' (Volkmann & Sievers 1979), a situation that itself confers special properties, if not functions, on the starch grains of statocytes.

(d) Certain maize mutants, 'amylomaize', have much smaller starch grains than the wild type and gravitropic sensitivity is also weaker in these mutants (Volkmann & Sievers 1979).

(e) Loss of gravitropism in decapped roots has already been mentioned. An important additional feature of this situation is the very close correspondence between the kinetics of gravitropic recovery and the behaviour of starch grains in the reforming cap (Wilkins 1984). Starch grains are, in fact, present before any gravitropic ability reappears, but the grains only become capable of sedimenting within the cells at about 12–24 hours after decapping; this is around the time at which gravitropism also reappears.

(f) In *Avena fatua* there is a very close correlation between an increase in gravitropic responsiveness and an increase in the number of starch grains per statocyte (Wright 1986).

However, there is also some circumstantial evidence against the starch grain × statolith correlations (Westing 1971, Pickard 1985a, c):

(a) Several types of green plants, from algae to some higher plants, have no starch grains yet show gravitropic responses (Westing 1971).

(b) Some mutants, both of *Zea* and *Arabidopsis*, contain no starch grains yet still respond gravitropically (Caspar *et al.* 1985).

(c) Fungi do not form starch, yet many have strongly gravitropic reproductive structures.

The absence of starch grains need not nullify the statolith concept, however. At least one case is known of non-starch statoliths in plants. The rhizoids of the sessile alga, *Chara*, grow by accretion of new wall material at the rhizoid tip, and are strongly orthogravitropic (Sievers & Volkmann 1979). Each rhizoid is a cylindrical cell of 30 μm diameter and several centimetres long, in which the contents show a strong ultrastructural polarity: about 10–20 μm behind the tip are 50 or so sacs which contain crystals of $BaSO_4$; further back from the tip, behind these $BaSO_4$ sacs, are the Golgi organelles that produce wall materials. Thus for wall growth, the Golgi vesicles must be transported past the $BaSO_4$ sacs to the rhizoid tip. When the rhizoid is horizontal, the sacs of $BaSO_4$ fall to the lower flank and somehow slow down wall growth. Relatively more growth therefore occurs on the upper side and the rhizoid curves downwards (Fig. 3.6). The specific mechanism by which the $BaSO_4$ statoliths inhibit growth is unknown. It has been suggested that they may interfere with the transport of the Golgi vesicles containing materials for wall synthesis (Sievers & Volkmann 1979) or that they may interact with a mechanoreceptive pressure-sensitive structure (Friedrich & Hertel 1973).

a: 0 min

b: 3 min

c: 6 min

d: 10 min

e: 15 min

f: 25 min

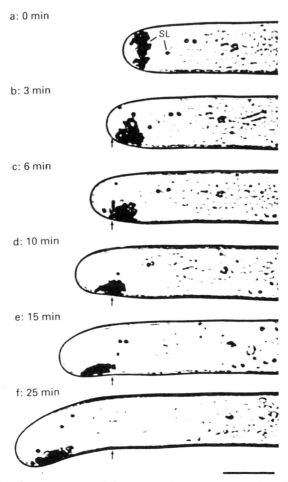

Figure 3.6 Time-lapse sequence of the gravitropic response in a rhizoid of *Chara*; SL, statolith; the arrow indicates the same point in the wall in each micrograph (from Sievers & Volkmann 1979).

In higher plants, too, the means by which the statoliths create the physiological asymmetry that underlies gravitropic curvature is unknown. Originally, statoliths were considered to move within the cell and thus interact with some other cell component, say a cell membrane (in which case the true gravity sensor would be the statolith–membrane interaction). Various possibilities and refinements have been suggested:

(a) Interaction of the statoliths with cell membranes may result in blockage of the plasmadesmata channels, and thus create a lateral asymmetry of transport capability (Juniper 1976).

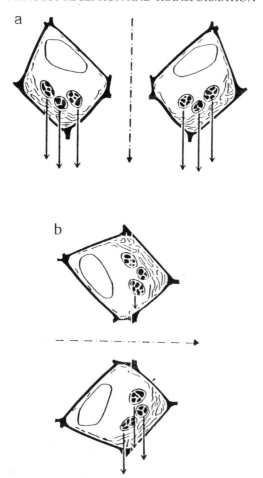

Figure 3.7 Events in reoriented root statocytes, according to the 'pressure model' of Volkmann & Sievers; the dashed arrows point to the root tip. (a) In a vertical root the pressure exerted on the ER by the amyloplasts is equal in statocytes on either side of the root. (b) In a horizontal root the pressures are unequal in statocytes on opposite sides of the root. (From Volkmann & Sievers 1979.)

(b) Interaction of the statoliths with the plasmalemma may result in the activation of enzymes involved in the release of hormones from some bound state (Kaufman *et al.* 1987). (However, although starch grains have been observed in association with the plasmalemma in cells of grass nodes (Parker 1979), this has never been seen in root tissues (Audus 1979, Moore & Evans 1986), even after considerable periods of gravitropic stimulation.)

(c) Interaction of the statoliths with the endoplasmic reticulum (ER) may result in the release of calcium, with consequential effects on hormone transport or action (Pickard 1985c, Moore & Evans 1986).

(d) A refinement of the statolith–ER interaction model takes note of the fact that at least some events in gravity sensing seem capable of occurring within less than a second, and suggests that statoliths may not actually have to move in order to initiate events (Volkmann & Sievers 1979); for example, consider two statocytes on either side of the root cap, in which the statoliths normally lie closely appressed on the ER (Fig. 3.7a); when the whole organ is turned on its side (Fig. 3.7b), and without any actual movement of the statoliths, there will be immediate changes in the pressures exerted by the statoliths on their respective ER, and a consequent difference between the two statocytes on either side of the root cap. Such a pressure difference may be the first step in establishing some physiological asymmetry across the organ. It has been further suggested that the cytoskeleton may also be involved in gravitropism, in integrating the function of ER and statoliths (Hensel 1984). [However, it should be borne in mind that not all statocytes have a layered arrangement of ER at the distal end of the cell, and not all starch grains seem to be in contact with the ER (Moore & Evans 1986). In fact, the vacuolated statocytes of the grass node (Parker 1979) seem to have an unusually sparse content of ER.]

Certain concepts that are being developed from studies in another area of investigation, however, may also prove useful in elucidating events in gravitropic stimulation. These are concerned with the possible presence of 'stretch-activated' ion gates in the cell membrane. Such channels are thought to be a means by which mechanoreception can occur in some animal cells (Guhary & Sachs 1984) and ciliated organisms (Taylor & Panasenko 1984). They have also been suggested to be involved in the reception both of mechanical stimuli (see Ch. 5) and of a gravity stimulus by plant cells (Edwards & Pickard 1987, Bjorkman 1988). That is, gravity is considered to have differential effects on the plasmalemma, opening some ion channels and closing others; statoliths may act as a means of focusing the effects of gravitationally induced mechanical stress onto the membrane (Edwards & Pickard 1987). The consequent ion flux may then form the basis of the physiological asymmetry across the tissues (and, as discussed in the following sections, may involve the differential transport of calcium and auxin, or the development of differential sensitivity of tissues towards those substances).

The evidence for mechanostimulation of the membrane being involved in gravitropic responses is presently of a very preliminary nature. In the first place, transient changes in electrical activity, indicative of transient ionic fluxes, have been recorded in gravitropically stimulated organs; these are discussed in the next section. Secondly, it has already been pointed out (Section 3.1.4) that forms of mechanical stimulation such as contact, friction, or flexure generally modify or override the gravitropic system, which might be expected if the two sensory systems shared common aspects of stimulus reception (Edwards & Pickard 1987). And thirdly, mechanical deformation of the cell membrane does seem capable of bringing about growth responses. In *Phycomyces* sporangiophores it has been demonstrated that longitudinal compressive stress enhances the growth rate, whereas tensile stress, or stretching, inhibits growth (Dennison & Roth 1976, Dennison 1971). And in dandelion peduncles it has also been shown that laterally applied compressive and tensile stresses induce growth responses that are commensurate with a gravitropic response in a shoot, i.e. when the organ is mechanically bent over towards one side it subsequently tends to grow away from that side (Clifford *et al.* 1982).

3.2.3 *Physiological mediation of the stimulus*

Electrical responses

Around the beginning of this century it was observed that a significant difference in electrical potential developed across a plant organ whose orientation had been altered, and this became known as the 'geo-electric effect'. It was also found to occur in inert material, such as salt-soaked paper, and at one time this seeming differential diffusion of ions was thought to offer important clues to the mechanism of gravitropism. However, in the 1960s it was clearly demonstrated that this geo-electric effect in plant organs was, in fact, a composite electrical response that was made up of an immediate but artefactual effect on the measuring electrodes, and a later biological effect (see Wilkins 1984, Pickard 1985a). The biological effect took around 15–20 minutes to develop, and was therefore considered to be a result of events involved in gravitropic curvature rather than a cause of them. However, more recent measurements have shown that there are, indeed, changes in the electrical activity of plant organs during the very early stages of the gravitropic response.

It is now well established that there are characteristic patterns of electric current associated with growing plant organs, and that these play major roles in growth and development (Bentrup 1984). In roots the described patterns of baseline current differ somewhat among investigators (Behrens *et al.* 1982, Miller *et al.* 1986, Bjorkman & Leopold 1987a), perhaps due to differences between species or culture conditions. However, there is general agreement that changes in these patterns occur as a result of changes in root orientation. In cress roots the symmetrical pattern of inward flowing current around the root tip becomes asymmetrical within 30 seconds of the onset of gravistimulation, flowing acropetally on the upper side and basipetally on the lower side (Behrens *et al.* 1982). In slower-reacting maize roots, there is an abrupt change in density of the outward flowing current at 3–4 minutes after gravistimulation (Bjorkman & Leopold 1987a). This shift in current density in maize roots seems to correlate well both spatially and temporally with features of gravity-sensing: it is associated only with the region adjacent to the columella of the cap, and its timing is coincident with the rather long presentation time of around 4 minutes that is characteristic of this tissue.

The changes in current flow in cress roots have subsequently been correlated with electrical events at the cellular level (Sievers *et al.* 1984, Behrens *et al.* 1985). The statocytes on what becomes the lower side of the root depolarize within 8 seconds of a change in orientation. (This is at present the most rapid plant response to gravity stimulation that has been recorded.) The statocytes on the upper side of the root tip become hyperpolarized, and this results in a potential difference across the root and the asymmetric current flow. It has been suggested that these changes in the membrane potentials of the statocytes may be caused by changes in the cellular concentrations of calcium, perhaps resulting from pressure interactions of starch grains with calcium pumps on the membranes of the ER (Sievers *et al.* 1984, Behrens *et al.* 1985) or on the plasmalemma (Hepler & Wayne 1985, Bjorkman & Leopold 1987a, b).

Transverse differences in the electrical potential have also been recorded in gravitropically stimulated hypocotyls, with the lower side becoming relatively positive (Tanada & Vinten-Johansen 1980). These changes have been observed within a minute of stimulation, but continue to develop over periods of 10 minutes or even longer. It has therefore been suggested that, rather than being part of the events involved in the reception and transformation of the gravity stimulus, these changes are more likely to be the result of some subsequent part of the physiological chain of events (Pickard 1985a, Behrens *et al.* 1985).

Calcium distribution

The suggestion of a possible involvement of calcium in gravitropism derives from early studies which showed that relatively large amounts of this ion are present on the upper side of gravitropically stimulated hypocotyls (Arslan-Cerim 1966, also Bode 1959 cited in Pickard 1985a). These studies were later extended (Goswami & Audus 1976), and calcium was observed to accumulate on the potentially concave side of tropically stimulated hypocotyls before there was any detectable curvature. (These latter authors, however, concluded that calcium redistribution did not play any causal role in the development of curvature, since unilateral treatment with mersalyl created a transverse asymmetry in calcium concentration but did not cause curvature.)

The asymmetry in calcium concentration in gravitropically stimulated shoots seems to be due largely to calcium localized in the cell wall (Slocum & Roux 1983). Nevertheless, a causal involvement of calcium redistribution in the gravitropic response has been concluded from the findings that it takes place within a few minutes of gravistimulation, before any curvature is apparent (Slocum & Roux 1983), and that both the calcium redistribution and tropic curvature, but not growth, are inhibited by treatment with the calcium chelator EGTA (Daye *et al.* 1984). The functional significance of the asymmetry of calcium concentration is not clear. It may itself have regulatory effects on growth, or it may be associated with effects of auxin transport or action. The full range of possibilities is extensively discussed in Pickard (1985a). Note, however, that in shoots the gravitropically induced calcium gradient is upwards across the organ, i.e. in the opposite direction to the traditional auxin gradient.

In roots too, several lines of evidence support the notion that calcium is involved in the gravitropic response. Maize roots cultured in EGTA continue to grow but show no response to gravity until the chelator is removed and calcium added back to the medium (Lee *et al.* 1983). If calcium, albeit at the relatively high concentrations of 10 mM, is applied unilaterally in agar blocks to vertical roots, the roots curve towards the calcium-treated side. And gravitropic curvature in roots is also associated with an increase in calcium on the slower growing side, in this case the lower side. (Thus in roots, the downward transverse gradient in calcium is in the same direction as the traditional auxin gradient.) It is not clear how the increased concentration of calcium on the lower side is brought about. Treatment of roots with auxin transport inhibitors prevents the increase in calcium on the lower side (Lee *et al.* 1984), suggesting that calcium movement may be closely linked to auxin movement. On the other hand, it has been suggested that calcium may be transported across the root cap in the apoplast or in the mucilage on the root surface (Moore & Evans 1986). Alternatively, calcium may be released from intracellular stores in the ER, as a result of statolith–ER interaction (Sievers *et al.* 1984). Other studies which also link calcium to gravitropism in roots concern the calcium-modulated regulator protein, calmodulin (Bjorkman & Leopold 1987b, and references therein). There are high concentrations of calmodulin in the cells of the columella of the root cap. Treatment with red light brings about a parallel increase in calmodulin activity and geosensitivity, and inhibitors of calmodulin correspondingly inhibit gravitropic curvature and the changes in electric current density which seem to be associated with gravity-sensing in maize roots.

On the other hand, some workers consider the transverse asymmetries in calcium concentration to be a secondary event rather than a causal factor in the physiological mediation of the gravity stimulus (Goswami & Audus 1976, Migliaccio & Galston 1987). For example, in gravitropically stimulated epicotyls of pea, the development of a calcium asymmetry was prevented by inhibitors of auxin transport but was *un*affected by agents which specifically inhibit the transport and uptake of calcium itself. These findings prompted the suggestion that the asymmetry resulted from the displacement of calcium from the cell walls by an auxin-induced proton flux (Migliaccio & Galston 1987). Furthermore, it is worth bearing in mind that the cytosol concentrations of calcium are around 10^{-7} M and that its regulatory effects also occur around this level (Hepler & Wayne 1985), rather than at the millimolar levels studied in many of the gravitropic investigations.

3.3   REGULATION OF THE GROWTH RESPONSES

During tropic curvature of a multicellular organ, growth on the convex side occurs at a relatively faster rate than on the concave side. The traditional Cholodny–Went explanation for all such asymmetries in growth is asymmetry in auxin concentration brought about by the lateral transport of auxin across the organ (see Ch. 2). Thus the effects of gravity cause more auxin to be present on the lower side of an organ. In shoots this brings about growth stimulation on the lower side (and upward curvature), whereas in roots, because of their greater sensitivity to auxin, it results in growth inhibition on the lower side (and downward curvature).

This explanation for the regulation of the growth events underlying gravitropic curvature has, until recent years, been tested mainly by the obvious approach of measuring the amounts of auxin in the upper and lower halves of gravitropically stimulated organs (see Audus 1979, Wilkins 1984, Pickard 1985a). However, the results of auxin analyses are not unequivocal (see later). Furthermore, the growth differential across a curving organ can be achieved in a number of ways other than simply by growth stimulation of one side or growth inhibition of the other (Firn & Digby 1980, Firn 1986a). And in fact, some of the recent observations of the manner in which these growth responses come about have resulted in suggestions that they cannot be regulated by the actual movement of auxin.

These aspects are considered in this section.

3.3.1   *Roots*

Growth responses
The growth responses underlying tropic curvatures in roots are confined to the region a few millimetres behind the root tip. In horizontal maize roots, downward curvature has been implied to result from differential

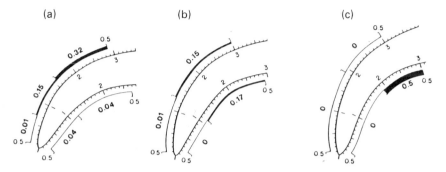

Figure 3.8 Diagrammatic representation of growth rates in different regions of cress roots at different times during gravitropic curvature; (a) 1.2 h, (b) 1.8 h, (c) 2.0 h. Numbers inside the root outlines are distances (mm) from the root tip; numbers outside the outlines are relative extension rates (h⁻¹), which are also represented by the thicknesses of the outer lines. (From Selker & Sievers 1987.)

inhibition of growth in which the rate on the upper side is reduced to a lesser extent than that on the lower side (Evans *et al.* 1986). However, considerable growth stimulation of the upper side has been observed frequently (Barlow & Rathfelder 1985).

Detailed measurements of growth rates in gravitropically responding cress roots indicate that curvature results from a composite series of responses that involve stimulation and inhibition of growth on each side of the root (Selker & Sievers 1987). In these studies three sequential phases of curvature were recognized, each with its characteristic pattern of growth (Fig. 3.8):

(1) Creation of curvature (0–1 h); growth on the lower side was inhibited, while on the upper side, in the region 1.58–2.84 mm from the tip, growth was stimulated to around three times faster than its previous rate.
(2) Maintenance of curvature (1–2 h); the growth rates were equal on both sides and the root continued to curve towards the vertical.
(3) Cessation of curvature (2 h); the root reached the vertical and curvature ceased; the growth rate fell to zero on the upper side while growth was initiated on the lower side in the region 2–3 mm from the tip.

(The timing of the phases and the actual growth rates are presumably related to the growth conditions and the temperature, which in the case described was 24 °C.)

These studies illustrate several points that should be borne in mind during any considerations of regulatory mechanisms. First, different parts of the root participate in the response, to different extents and at

75

different times. This spatial and temporal complexity of the response was manifest in the case of the cress roots in various ways; for example, the physical shape of the gravitropic curve was not uniform but consisted of two areas of curvature separated by a flatter region; again, growth stimulation occurred on each side of the root not only at different times, but also in different regions (towards the apex of the curve on the upper side, and at the base of the curve on the lower side). A second point that should be borne in mind is that for curvature to cease, the direction of the initial gradient of differential growth across the organ (faster on the upper side) must be *reversed* (faster on the lower side); if the growth rates on each side were only equalized (stage 2), curvature would simply continue into some kind of coiled configuration.

Regulation of response

In root gravitropism, the clear distinction between the region of stimulus reception in the cap, and the region of the growth responses at least a few millimetres behind the tip means that there must be some form of message transmission between these regions. Furthermore, since the surgical removal of the cap generally results in an increased rate of root growth (Juniper 1976), it seems reasonable to infer that this message is likely to be in the form of an inhibitor. The notion that a growth inhibitor originates from the root cap receives support from the results of experiments in which only half the cap is removed and the root curves towards the side with the remaining half cap, i.e. growth is inhibited on that side (Wilkins 1984, Moore & Evans 1986).

The nature of this inhibitor, however, is not clear. Abscisic acid (ABA) is present in roots and its synthesis is stimulated by light (Wilkins 1984), a treatment that also enhances gravitropic sensitivity in many species (see Section 3.1.4). But several lines of evidence indicate that ABA is unlikely to be the tropic growth regulator. First, unilateral application of exogenous ABA does not bring about curvature (Mertens & Weiler 1983). Secondly, ABA gradients are generally not found in tropically stimulated roots, either of endogenous ABA or of exogenously supplied radiolabelled ABA (Jackson & Barlow 1981, Mertens & Weiler 1983); an exception to this second generalization is the report of Pilet & Rivier (1981). And thirdly, gravitropism still occurs even under conditions in which the possibility of actually forming a gradient of ABA is made highly unlikely; for example gravitropic curvature still occurs in the roots of plants in which the synthesis of ABA is inhibited either by chemicals or by mutation; and it also still occurs when roots are bathed in an external solution of ABA (Moore & Evans 1986).

Another candidate for the role of tropic growth regulator may be a second inhibitor that also seems to be present in the root cap (Suzuki *et al.*

1979). This compound is known not to be ABA but otherwise is still unidentified.

The regulator originally associated with the inhibition of root growth is, of course, auxin, and many investigators have returned to considering this to be the tropic growth regulator in roots (Moore & Evans 1986). In this respect, the asymmetric application of exogenous IAA to decapped roots does at least result in curvature of the root towards the side to which the IAA is applied (Moore & Evans 1986). However, although there are a few reports of greater concentrations of auxin on the lower side of gravitropically responding roots (see Jackson & Barlow 1981) this aspect is by no means clearly established (Mertens & Weiler 1983).

The mechanism by which any auxin gradient could be created has also not been clarified. Endogenous auxin transport in roots is highly polarized in the direction of the root tip, and occurs largely in the stele. It may then be transported by some active, non-diffusional process back along the root in the non-vascular tissues to regulate growth in the elongation zone (Wilkins 1984). It has been suggested that the asymmetric distribution of calcium to the lower side of the root cap (see Section 3.2.3) may bring about a greater lateral distribution of auxin to the lower side and a consequently greater transport of auxin along the lower side to exert inhibitory effects on the cells of the elongation zone (Evans et al. 1986, Moore & Evans 1986). Alternatively, or additionally, the greater amounts of calcium on the lower side of the root may enhance the longitudinal transport of auxin on that side of the root (Pickard 1985a).

However, a recent study has shown that the gravitropic response is not eliminated even when the epidermis and outer cortex are stripped from a short length of root behind the apex (Bjorkman & Cleland 1988). This suggests that any movement of auxin, and of any other gravitropic signal, occurs in the inner tissues rather than the cortical tissues.

3.3.2   Shoots

Grass stems

The false pulvinus, or 'node', of the grass stem represents a situation that is anatomically self-contained and that shows a localized gravitropic response in which the growth responses are well established (Wilkins 1984, Kaufman et al. 1987). Upward bending in a horizontal node is due to a graded initiation and promotion of growth across the organ. The same form of gravitropic response is also shown by isolated nodes. However, even though reception of the gravity stimulus and development of the growth response must therefore both take place within the same organ, there may still be a need for some form of physiological message transmission. The gravity-sensing cells, i.e. those containing starch-grain statoliths, are several cells distant from the starch-free cells

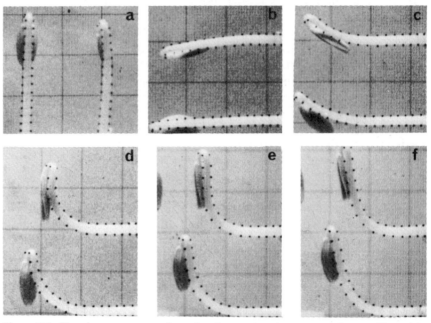

Figure 3.9 Time-lapse sequence of gravitropic curvature in two cucumber seedlings. (a) Marker beads enable the extent of curvature and the rates of growth in different regions to be determined, the background grid lines are 1 cm apart; (b) 0 hr; (c) 1 h; (d) 2 h; (e) 3 h, (f) 4 h. (From MacDonald *et al.* 1983b.)

of the collenchyma whose elongation forms the basis of the gravitropic growth response (Kaufman *et al.* 1987). The question is whether the movement of auxin is the basis of this message transmission.

In isolated grass nodes, application of exogenous auxin generally stimulates growth, and unilateral application also induced a bending response whose form and kinetics were remarkably similar to that induced by gravistimulation (Clifford 1988). Appropriate asymmetries in auxin concentration, with more auxin on the lower side, are found after gravistimulation (Wright *et al.* 1978). However, a gravitropic response still occurs in the isolated lower halves of longitudinally bisected nodes, arguing against the involvement of lateral transport of auxin right across the organ, and in fact gravistimulation did not induce any lateral transport of exogenously applied IAA in wheat nodes (Bridges & Wilkins 1973). It has therefore been suggested that the auxin asymmetry arises from a combination of increased synthesis of auxin together with the release of auxin from some bound form (Kaufman *et al.* 1987). Comparison of the lags in the response times to stimulation by gravity and by uni-lateral application of IAA indicates that there would be several minutes available for such asymmetries to develop (Clifford 1988). However, it

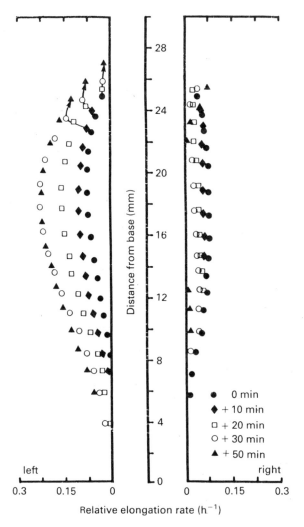

Figure 3.10 Growth rates along each side of a horizontal hypocotyl of sunflower during gravitropic curvature; growth rates are expressed as relative elongation rates (h⁻¹); the left-hand side is the lower side (from Berg *et al.* 1986).

has also been suggested that changes in tissue sensitivity to auxin may be involved in the gravitropic responses of the grass node (Clifford 1988).

Growth responses in coleoptiles and other shoots

Initial studies of the growth responses of gravitropically curving hypocotyls (Digby & Firn 1979) suggested that growth inhibition of the concave side of the organ was the major, and often the only, change in growth responsible for curvature (thus making unnecessary any require-

Table 3.1 Growth rates (μm h^{-1}) of successive zones from apex (zone 1) to base (zone 12) in an etiolated cucumber hypocotyl undergoing gravitropic curvature in darkness; the values at 0 h are the growth rates in the vertical seedling for the hour preceding gravistimulation; negative values are zones which actually decreased in size during that hour, a phenomenon documented by several authors (data from MacDonald *et al.* 1983b).

	Lower side					Upper side				
Zone	0 h	1 h	2 h	3 h	4 h	0 h	1 h	2 h	3 h	4 h
1	275	242	220	62	308	110	176	154	194	66
2	110	352	176	− 26	154	176	88	44	422	110
3	154	286	242	− 36	66	143	− 44	66	294	198
4	66	176	286	44	− 132	132	0	0	168	286
5	198	352	242	62	154	110	− 22	− 110	194	264
6	121	132	264	123	88	132	22	− 44	52	154
7	99	198	264	185	− 44	99	− 66	0	− 27	88
8	88	176	220	338	0	110	− 22	− 44	− 92	0
9	77	176	66	194	132	88	22	− 132	82	44
10	11	44	154	− 22	154	22	− 22	0	− 233	176
11	66	22	22	88	44	77	− 22	0	66	− 154
12	0	− 22	22	66	− 22	5	66	0	0	0

ment for, or involvement of, auxin-stimulated growth). And they further suggested (Firn & Digby 1980) that the changes in growth responsible for curvature occurred simultaneously and uniformly along the length of the organ (thus making unnecessary any requirement for longitudinal transmission of some physiological message).

The growth events underlying gravitropism have since been investigated in a number of organs. Methods have generally involved making time-lapse films or video-recordings of organs that are undergoing curvature, with their surfaces marked in some way, say by application of ink marks, lanolin 'hairs', or ion-exchange beads (Fig. 3.9). Subsequent frame-by-frame measurements of the changing positions of these markers can then be processed to give the relative elemental elongation rates along each side of the responding organ (Fig. 3.10). (The mathematical 'smoothing' procedures, however, tend to minimize the very marked differences in growth rate that can exist in adjacent regions of an organ, cf. the data in Table 3.1.)

There is now general agreement that gravitropic curvature in shoots, both in seedlings (MacDonald *et al.* 1983b, Bandurski *et al.* 1984, Jaffe *et al.* 1985) and in more mature plants (Mueller *et al.* 1984), results from stimulation of growth on the lower side of an organ and inhibition of growth on the upper side. It has been further suggested that, at least in seedlings, the gravitropic response can be considered as involving two

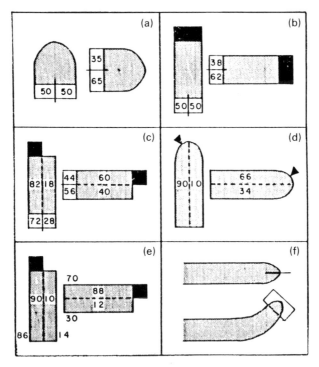

Figure 3.11 Summary of various types of experiments to determine whether lateral transport of auxin occurs in the coleoptile during gravitropism; coleoptiles are shaded, receiver agar blocks are white, donor blocks (or spots) are black; solid lines represent impermeable barriers; numbers indicate the percentage distribution of IAA in blocks or tissues. (a) Bioassays of Dolk in 1929; (b) radioassays of Gillespie & Thimann in 1961; (c) asymmetric applications of IAA by Goldsmith & Wilkins in 1964; (d) spot applications of IAA by Shaw *et al.* in 1973; (e) asymmetric applications of IAA by Wilkins & Whyte in 1968; (f) barrier experiments of Brauner & Appel in 1960. (From Wilkins 1984, in which the original references are cited.)

phases (MacDonald & Hart 1987b); the initial phase consists of rapidly imposed growth inhibition and growth stimulation on opposite sides of the hypocotyl within 5–10 minutes of stimulation; the later phase, over the next few hours, only occurs if the apex of the organ is present, and consists of the basipetal migration of growth stimulation along the lower side of the organ right into the previously non-growing basal region itself. (This effect can be seen both in the marked basal zones of the hypocotyl in Fig. 3.9 and in the growth data of Fig. 3.10.) Overlying the later phase of the response are the growth changes associated with the autotropic straightening of the organ; these autotropic changes also involve growth stimulation and inhibition, and again develop downwards from the apex (Firn & Digby 1979, MacDonald *et al.* 1983b). Thus, the complete attainment of a vertical position by a shoot involves a fairly complex sequence of changes in growth rate with, in the later stages, stimulation *and* inhibition of growth occurring simultaneously on the same side of the organ (Table 3.1).

Involvement of an auxin gradient

Some of the key findings in regard to the involvement of an auxin gradient in shoot gravitropism are summarized diagrammatically in Figure 3.11. These include demonstrations, both by bioassay (Fig. 3.11a) and by radioassay methods (Fig. 3.11b), of greater amounts of auxin in the lower halves of gravitropically stimulated coleoptiles. Many subsequent investigations have confirmed such asymmetries in auxin concentration both in coleoptiles and in dicot stems (documented in Pickard 1985a, c).

Other studies have attempted to determine where in the organ, and how, such asymmetries become established. The results of early barrier experiments indicated that lateral transport of auxin occurred in the tip of a horizontal organ (Fig. 3.11f). [However, it has been pointed out (Firn *et al*. 1981) that these early experiments took no account of the anatomical asymmetry of the coleoptile, and that the results could have arisen from damage to the vascular bundles, rather than from an inhibition of lateral transport.]

In later experiments, radiolabelled auxin was asymmetrically applied to the tissues, and the results suggested that lateral transport also occurred across the body of the organ (Fig. 3.11c, d, e). Enhanced longitudinal transport in the lower half of a horizontal organ has also been implicated (Wilkins 1984). [The interpretation of these later experiments, and distinction between gravitropically stimulated lateral transport and enhanced longitudinal transport are not as straightforward as may at first appear; see Pickard (1985a) for a thorough discussion of this problem.]

However, despite the large volume of work on the subject, the analyses of auxin asymmetries are not unequivocal, and this has led to the expression of serious doubts about the overall validity of the Cholodny–Went explanation of tropic curvatures (Firn & Digby 1980, Firn & Myers 1987). These doubts relate to several aspects of the theory.

It has been repeatedly demonstrated that, unlike the situation in the root, the presence of the apex is not necessary for a gravitropic response in shoots. Therefore the notion that the apex is the specific site of stimulus reception has been questioned (Firn *et al*. 1981). This has been countered by the argument that the facility for lateral transport of auxin exists throughout the length of the organ (Pickard 1985a). Furthermore, the apex seems to play some role in the response. It has been shown that for the full expression of the gravitropic response, including in particular the basipetal migration of growth stimulation along the lower side of a hypocotyl (cf. Figs. 3.9 & 3.10), the apex must not only be present but it must also be undergoing gravitropic stimulation (Hart & MacDonald 1984, MacDonald & Hart 1985a, b). The importance of the orientation of the coleoptile apex was more dramatically demonstrated by Francis Darwin (1899). Using seedlings of *Phalaris, Sorghum,* and *Setaria,* he showed that when the apex of the coleoptile was held in a horizontal orientation (by partial insertion into a glass tube), the seedling took up a coiled configuration, largely through the continuing (tropic) growth of the mesocotyl.

Another area of doubt about the Cholodny–Went theory relates to the actual occurrence of lateral gradients of auxin across an organ. The finding that the isolated lower half of a longitudinally split shoot can still respond gravitropically, i.e. can still respond in the absence of any upper half from which growth regulator could be transported, is often put forward as evidence against the idea that an auxin gradient can be involved in tropic curvature. In fact, this finding led to Gradmann (1925) proposing his 'co-factor theory' of gravitropic curvature, in which the growth differential across a horizontal organ was conjectured to arise, not from a change in auxin concentration, but from a gravitropically induced change in some co-factor that affected the responsiveness of the tissues towards auxin (a seminal notion, perhaps, of gravitropically induced change in tissue sensitivity to auxin). Gradmann's specific proposals for an auxin co-factor have since been abandoned, but the finding of gravitropic behaviour in the isolated lower half of an organ remains. Its explanation may lie in the relatively greater role played by the epidermal tissues in growth regulation, and this aspect is discussed later.

Doubts have also been expressed as to whether auxin gradients that have been detected in gravitropically responding tissues are both large enough and formed quickly enough to account for the growth responses. With regard to the magnitude of the gradients, it has been pointed out, reasonably enough, that little can be said on this matter until it is known whether auxin is distributed uniformly or not throughout the tissues of each half of the organ (Pickard 1985a). This counterargument receives more support from the evidence that it is the behaviour of the epidermal tissues in particular that is important in the growth of an organ (see later). Nor do the kinetics of auxin gradient formation seem to present much of a problem. Auxin gradients have been detected fairly rapidly after gravistimulation, within 5–10 minutes in coleoptiles (Filner & Hertel 1970) and in dicot stems (Harrison & Pickard 1986b) and even within 3 minutes in maize mesocotyls (Schulze & Bandurski 1986, 1987). And again, if it is auxin movement between particular tissues that is important, rather than across the bulk of the organ, kinetic difficulties are further eased.

Thus go the arguments and counterarguments about the involvement of an auxin gradient in tropic curvature, possibly never to be completely resolved to the satisfaction of all protagonists. However, there is no question that the original Cholodny–Went concepts need further examination. For example, gravitropic curvature can clearly occur in the absence of detectable auxin gradients (or of any other hormonal gradients), even when highly sensitive immunoassay methods are used to measure hormonal concentrations (Mertens & Weiler 1983). Secondly, and perhaps less contentiously, it has to be questioned whether simple auxin gradients alone can account for the patterns of growth now known to underlie tropic curvature in shoots and in roots, in particular, patterns that involve growth stimulation and inhibition simultaneously on the same side of an organ and that involve reversal of the growth gradient across an organ within a single region of tropic curvature.

Other hormones
There are no clear indications of the general involvement of any of the other plant growth regulators as primary mediators of the growth responses in shoot gravitropism. Transverse gradients of gibberellin have been observed in gravitropically responding hypocotyls of sunflower (Phillips 1972a), but these were detected at relatively late stages of curvature. Furthermore, unilateral application of exogenous gibberellin does not usually result in curvature of an organ, nor does a bathing solution of gibberellin prevent gravitropic curvature (MacDonald & Hart 1987b). However, spray applications of GA_3 brought about upward bending in the plagiotropic rhizomes of Bermuda grass (Montaldi 1969), and asymmetric distributions of gibberellin activity have been detected in grass stems, albeit after very long periods of gravistimulation (see Pickard 1985a).

Asymmetric distributions of an ABA-like compound have been detected in a 'weeping' variety of mulberry (Reches *et al.* 1974). In this case, however, the greater concentrations of the compound were found on the *upper* side of the downward growing branches, the significance of which is unclear.

It has been suggested that ethylene may be directly involved in shoot gravitropism (Wheeler & Salisbury 1981). However, these studies involved the use of ethylene inhibitors and the specificity of the effects, and their particular relationship to gravitropism, have been questioned (Harrison & Pickard 1986a). The possibility is allowed, though, that ethylene may be involved in the growth counter-reaction that occurs in the later stages of gravitropism (Pickard 1985a). A primary role for ethylene in the gravitropic responses of grass nodes has also been discounted, since increased synthesis of ethylene is not observed until 5.5 hours after gravistimulation (Kaufman *et al.* 1987).

Models for the regulation of gravitropic curvature

One model ascribes a nutritional rather than hormonal basis to tropic growth, by proposing that the differential growth is due to the differential availability of water within the tissues (McIntyre 1980). Evidence is presented in support of the model that in gravitropically stimulated hypocotyls the cells on the upper side have a lower water content, even when the tropically stimulated hypocotyl is physically restrained from developing tropic curvature. (In phototropically stimulated organs the cells on the lit side also have a lower water content.) These are intriguing experimental findings, but recent analyses indicate that wall loosening, not cell turgor, is the limiting factor in growth at the cellular level (Cosgrove 1987).

Most investigators continue to ascribe some form of regulatory role to auxin. It is well established in a variety of organs that curvatures develop specifically in response to the unilateral application of IAA (Ullrich 1978, Clifford *et al.* 1985, Clifford 1988). (In many of these cases, there is also an initial phase of 'wrong-way curvature' in which the organ transiently bends towards the side of IAA application, analogous to the early wrong-way gravitropic curvature often seen in roots and coleoptiles.) Furthermore, when hypocotyls are bathed in a solution of IAA, which presumably nullifies any endogenous differentials in auxin concentration, they show no graviresponse whatever, although they continue to grow normally (MacDonald & Hart 1987b).

Several recent accounts of the possible regulatory background to gravitropism invoke some kind of interaction between auxin and calcium. Hertel (1983) links auxin action with auxin transport and calcium uptake. He suggests that the plasmalemma carries an electrogenic symport for the influx of auxin and an anion carrier for its efflux, coupled to calcium influx. He further suggests that any 'stimulus' (such as gravity) affects the net influx of auxin and calcium by affecting the properties of the anion carrier. Therefore, the presence of auxin increases the calcium concentration of a cell, and also stimulates calcium movement. Pickard (1985a) develops further the notion of coupled auxin and calcium transport and proposes that calcium inhibits growth, either directly or indirectly by activating the transport of auxin to the gravitationally lower side of an organ. She also widens the area of stimulus reception to include not only the action of gravity on calcium channels in the membrane, either directly or mediated through statoliths, but also the action of mechanical stimulation, and hence accounts for the effects of such factors as wind flexure and soil friction in overriding responses to gravity.

It is becoming clear that the shoot epidermis plays a major regulatory role in controlling growth (see Box 3.2). It has also been known since the work of Sachs and of Copeland around 1900, on the sluggish responses of

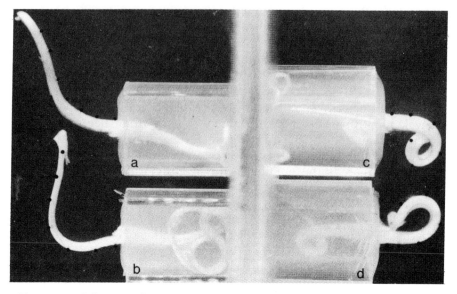

Figure 3.12 Behaviour of longitudinally split hypocotyls of sunflower in solutions of buffer or IAA (10 μM). (a), (b) 'Upper' and 'lower' halves in buffer; both show some gravitropic curvature. (c), (d) 'Upper' and 'lower' halves in IAA; the IAA-stimulated growth of the epidermis brings about 'inward' coiling in both orientations. (From MacDonald & Hart 1987a, b.)

peeled stems and hypocotyls, that the epidermis is important in gravitropism (Pope 1982), and regulatory models of gravitropic curvature have been proposed in which the epidermis plays a central role (Iwami & Masuda 1976, Firn & Digby 1977). Both of these models, however, seem still to envisage the regulation of differential growth to involve the lateral movement of auxin completely across the responding organ from the upper to the lower epidermis.

A more recent model adds further refinement to the roles of the peripheral tissues (MacDonald & Hart 1987a, b). Central to this model is the fact that the epidermal and sub-epidermal tissues are differentially sensitive to auxin. A simple illustration of this differential sensitivity is seen in the effects of auxin on longitudinally split hypocotyls (Fig. 3.12). Due to auxin stimulating the growth of the epidermal cells, but not that of the inner sub-epidermal cells, each half of the hypocotyl curves inwards (no matter what its orientation) and eventually forms coils. In fact, this response formed the basis of Went's original split pea-stem bioassay for auxin, since in that tissue the amount of inward curvature is strictly proportional to the concentration of auxin. The effect is quite specific for auxin and has long been known to be due to the differential responsiveness of the outer and inner stem tissues to auxin (Thimann & Schneider 1938, Kutschera et al. 1987).

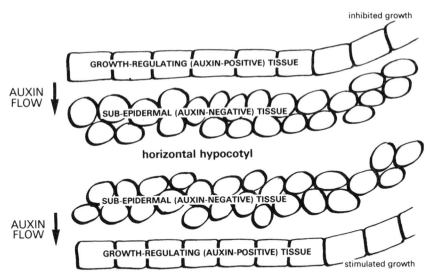

inhibited growth

GROWTH-REGULATING (AUXIN-POSITIVE) TISSUE

AUXIN FLOW

SUB-EPIDERMAL (AUXIN-NEGATIVE) TISSUE

horizontal hypocotyl

SUB-EPIDERMAL (AUXIN-NEGATIVE) TISSUE

AUXIN FLOW

GROWTH-REGULATING (AUXIN-POSITIVE) TISSUE

stimulated growth

Figure 3.13 Diagrammatic representation of possible early events during gravitropism in a dicot shoot; limited auxin movement between the epidermal and sub-epidermal tissues results in rapid growth inhibition of the upper side and stimulation of the lower side (from MacDonald & Hart 1978a, b.)

If this situation of differential tissue sensitivity is applied to the events of gravitropism (Fig. 3.13) then, when a shoot is turned on its side, auxin could move rapidly out of the epidermal cells of the upper side (making their auxin concentration sub-optimal for growth) and into the sub-epidermal cells of the upper side (which are insensitive to, or even inhibited by, auxin). Meanwhile on the lower side of the shoot, auxin would move out of the sub-epidermal cells and into the epidermal cells. The net result of this limited movement of auxin would be a rapid cessation of growth on the upper side and stimulation of growth on the lower side.

Realization of the markedly different auxin sensitivities of closely adjacent tissues puts an entirely new perspective on the extent of auxin movement that may be required to bring about large differentials in growth rate, i.e. virtually from one cell to another rather than from one side of the organ to the other. It is intriguing that an early investigation suggested that the auxin differential between the surface layers of the upper and lower sides of a horizontal coleoptile can be as great as 10:1, but that this is obscured in assays of half segments in which the bulk of the tissue does not show much change in auxin concentration (Burg & Burg 1967). These early results were obtained by extrapolating trends in the distribution of radioactivity across tissues which had been supplied with radiolabelled IAA. More radiolabelled IAA was also detected in the

lower epidermis than in the upper epidermis in gravitropically stimulated hypocotyls of cucumber (Iwami & Masuda 1976). On the other
hand, measurement by immunoassay techniques revealed no differences
in auxin content between the peripheral layers and the central tissues in
several species (Mertens & Weiler 1983). However, in these later studies
the peripheral layers consisted of 2–3 layers of cells. It is the differential
between the true epidermis and the immediately subjacent cells that is
suggested to be critical.

Other investigators suggest that actual transport of auxin need not
occur in order to initiate changes in growth. For example, a stimulus may
bring about changes in the amount of free (or 'active') auxin through
effects on the sub-cellular compartmentalization of auxin (Mertens &
Weiler 1983).

On the other hand, there have been several recent suggestions that
physiological regulation derives from changing tissue sensitivity to the
regulator, rather than from changing levels of regulator (Trewavas 1981,
1982, 1986, Firn 1986b, Weyers et al. 1987). Such a mechanism has also
been specifically suggested to be involved in the regulation of gravitropic
responses (Clifford et al. 1985, Moore & Evans 1986, Salisbury et al. 1986,
Meudt 1987, Clifford 1988), and it may well be that mechanical effects of
gravity stimulation on the plasmalemma of the epidermal cells bring
about changes in their responsiveness to auxin.

In any event, it would seem that future investigation of the
mechanisms underlying gravitropism must be at the tissue and cell level,
rather than at the level of whole organs.

Box 3.2   The role of the shoot epidermis

Perhaps a popular notion of the shoot epidermis is that it is simply a passive outer covering, showing, at most, some protective adaptations against particular climatic or biological factors in the environment. However, it plays a much more active part than that in the general growth and development of the plant, acting in regulatory and perhaps also in sensory roles.

The growth regulatory role of the epidermis was first surmised over 100 years ago. When a non-woody stem or petiole is split longitudinally, the separate halves spring apart, and Sachs realized that this is due to the tensions generated by the different rates of growth of the different tissues. He further reasoned that, since the split sections curve outwards, the outer tissues must normally act as a restraining influence on the growth and expansions of the inner tissues. This notion of Sachs has been refined and extended by recent studies in which the growth of an organ is suggested to result from the interaction between the expansive force generated by the inner tissues and the constrictive force exerted by the epidermis (Kutschera *et al.* 1987, Kutschera & Briggs 1988). Furthermore, auxin has been shown to act exclusively on the extensibility of the cells of the epidermis, rather than on those of the central tissues (Brummell & Hall 1980, Kutschera & Briggs 1987, 1988). Thus the behaviour and responses of the epidermis exert a major regulatory influence on the overall growth of an organ. (But it is still not clear whether changes in growth rate are brought about by changes in the auxin content of the epidermis, or by changes in the auxin sensitivity of the epidermis.)

The epidermis may also play a major role in sensory responses. It has already been pointed out (Ch. 1) that temperature-induced differential growth responses of the upper and lower epidermis of the petals regulate opening and closing movements in flowers such as crocus and tulip. Further, the leaf epidermis is thought to be the site of stimulus reception in photoperiodic responses (Schwabe 1968, Mayer *et al.* 1973). It may also be involved in the specific directional reception of the light signal in some cases of heliotropism (Ch. 4). And the epidermis also seems to be the reception tissue for mechanostimuli (Ch. 5).

[A recent study in roots (Bjorkman & Cleland 1988) suggests that in these organs some inner cell layer, such as the endodermis, may be the growth regulatory tissue and the site of auxin action.]

CHAPTER FOUR

Phototropism

4.1   GENERAL INTRODUCTION

4.1.1   *Definition and general description*

Since green plants are directly dependent on light for their supply of energy, it is perhaps not surprising that the orientations of many of their organs are also markedly influenced by light. This influence is actually expressed through several different mechanisms, and the term photo-tropism is used to describe those situations in which orientation in relation to light is brought about by some form of direct growth response.

Phototropism is often defined as curvature of an organ in relation to the 'direction' of light, and certainly, in most laboratory investigations of the subject, responses occur in relation to the direction of some kind of unilateral illumination. But under natural conditions, the asymmetry in light more usually results from a non-homogeneous or unequal distribution of diffuse light. In either case, a gradient of light absorption is established across the organ and this results in some sort of physiological gradient which induces the appropriate growth pattern. [That curvature occurs in response to a light gradient, rather than to light direction, was elegantly demonstrated by the simple experiments of Büder in the 1920s (cited in Pohl & Russo 1984); he irradiated only one side of a coleoptile from above, and the coleoptile bent towards the irradiated side.]

Another feature that is popularly considered to be characteristic of phototropism is that it is brought about only by blue light. However, as we shall see, red light induces phototropic responses in several lower plants, and under certain circumstances it can also induce directional growth responses in the organs of higher plants, both directly and indirectly (see Sections 4.1.4 and 4.1.5).

In many higher plants light can also influence organ orientation through effects on turgor responses. Light-induced changes in turgor of the pulvinar cells are responsible for the photonastic movements and heliotropic behaviour of many leaves. These types of responses are also described in Section 4.1.5.

90

The treatment of the subject in this chapter follows more or less the same pattern as that for gravitropism in Chapter 3. However, it becomes apparent almost immediately that there are substantial differences in the relative sizes of analogous sub-sections, particularly in those of the general introductory section, 4.1. For example, phototropic responses seem much more widely investigated than gravitropic responses in lower plants (Section 4.1.2); light seems to exert a greater variety of directional effects than does gravity (Section 4.1.5); but, on the other hand, the phototropic responses themselves seem much less malleable and subject to change by other factors (Section 4.1.4), than do the gravitropic responses of an organ. These differences in emphasis between responses to light and gravity may result to some extent from the bias, or ignorance, of the author, but they are also a reflection of the overall importance of light for plant growth and the general dominance of light as a directional signal throughout the plant kingdom. This aspect is considered further in Section 4.4.

4.1.2   Responses in lower plants and fungi

In lower plants generally, light is a more commonly used and more dominant directional signal than gravity (Banbury 1959). Obviously, this is of adaptive value in the photosynthetic organs of green plants. Perhaps less obviously, non-photosynthetic organisms, such as fungi, also seem to utilize light as a more reliable signal than gravity for getting spores into dispersive air currents. (In the typical habitat of many fungi, dependence simply on the upwards and downwards signal of gravity could result in reproductive structures growing into blind cavities.)

There are two general means by which organisms can detect the direction or gradient of some environmental stimulus, known as 'temporal sensing' and 'spatial sensing' (see Section 2.1). In temporal light-sensing methods, the organism detects the level of light at two different points in time, and any change in light intensity initiates particular physiological responses. These kinds of methods are used by small, rapidly moving organisms, and they are responsible for the phototactic behaviour of such algae as *Euglena*, *Volvox*, and *Chlamydomonas* (Häder 1979, Feinleib 1980). In this form of sensing, it is the transient response to the change in the level of the stimulus that is crucial to the overall response, i.e. the phenomenon of sensory adaptation (Section 2.1) plays a key role. In spatial-sensing methods, differences in light intensity are detected in different regions of a cell or organ, and appropriate physiological responses initiated. It is this sort of method that is involved in the phototropic growth responses of large non-motile cells or multicellular organs, and the sensing accuracy is often refined by further mechanisms of attenuation (screening effects), refraction (lens effects), or dichroism

Table 4.1 Phototropic responses in lower plants and fungi (adapted mainly from information compiled by Pohl & Russo 1984).

Type of plant	Species and organ	Spectral region	Phototropic response	Type of growth	Reference
Algae	*Mougeotia* filament	blue & red	negative	bowing	Neuscheler-Wirth (1970)
	apical cell	red*	positive	bulging	
	Vaucheria filament	blue	positive	bulging	Kataoka (1975)
	Boergesenia rhizoid	blue	negative	bulging	Ishizawa & Wada (1979)
Mosses and liverworts	*Sphaerocarpus* germ tube	blue*	positive	bulging	Steiner (1969)
	Physcomitrium protonema	red	positive	bulging	Nebel (1968)
	Funaria protonema	red*	positive	bulging	Jaffe & Etzold (1965)
Ferns	*Dryopteris* protonema	blue* & red*	positive	bulging	Etzold (1965)
	rhizoid	red	negative	bulging?	Hartmann et al. (1965)
	Pteridium protonema	red	positive	bulging	Davis (1975)
Fungi	*Phycomyces* sporangiophore	blue	positive	bowing	Curry & Gruen (1959)
		UV	negative	bowing	Curry & Gruen (1957)
		λ 513–621 nm	positive	bowing	Löser & Shäfer (1980, 1986)
		blue, plus λ 414, 491, 650 nm			Galland (1983)
	Pilobolus sporangiophore	blue	positive	bulging (young) bowing (mature)	Page (1968)
	Entomophthora conidiophore	blue & red	positive	bowing	Page (1968)
	Coprinus stipe	blue	positive	bowing	Banbury (1959)

* Also show polarotropic responses perpendicular to the electrical vector of the light.

(differential light absorption by the photoreceptor). Again, sensory adaptation to light intensity is often apparent.

There are two general means by which the growth of organs can be directionally influenced by light, known as 'bulging' and 'bowing' (Section 2.1). 'Bulging' occurs in tip-growing, unicellular organs, and consists of the displacement of the growing point away from the tip, or the creation of a new growing point. A positive phototropic response by this method therefore involves stimulation of growth on the apical flank that is nearest the light. (Less commonly, negative phototropism by this method results from light inhibition of growth.) A 'bowing' response involves the establishment of differential rates of growth across some sub-apical growing region in an organ. The differential growth can arise from various combinations of differential stimulation or inhibition, but a positive phototropic response by this means involves a faster growth rate on the side of the organ that is farthest from the light source. Bowing forms of phototropic responses are seen not only in all multicellular organs, but also in some unicells which have a sub-apical growing zone.

A general outline of the phototropic responses in the various classes of lower plants is given below. The responses of the major organisms that have been studied in each class are summarized in Table 4.1. Details of action spectra, fluence responses, and growth behaviour are given in the reviews by Dennison (1979), Hertel (1980), and Pohl & Russo (1984).

Algae

Positive and negative phototropic responses are apparent in algae. There are also examples of polarotropism (see Section 4.1.5). In those algae in which there is some morphological differentiation, such as *Cladophora* and *Acetabularia*, the rhizoids are generally negatively phototropic whereas the fronds show positive or diaphototropic responses (Banbury 1959).

Phototropic responses have been particularly studied in certain filamentous algae. In *Mougeotia*, for example, light initiates several directional responses (Neuscheler-Wirth 1970): in the apical cell of a filament, red light stimulates growth and a positive phototropic response results from bulging on the lighted side; however, in the intercalary cells, both red and blue light stimulate growth and induce negative bowing responses. The phototactic movements of *Mougeotia* chloroplasts have also been extensively investigated (Haupt 1965, 1983): in low-intensity light (daylight), the chloroplast moves to face the light in response to a red-light-stimulated phytochrome reaction; in high-intensity light (sunlight), it turns protectively edge on to the light as a result of a blue-light-stimulated reaction.

In *Vaucheria*, blue light induces a shift in the growing point of the filament towards the light, to give a positive phototropic response

(Kataoka 1975, Kataoka & Weisenseel 1988). The chloroplasts of *Vaucheria* also show phototactic movements in response to light intensity, but in this case the same blue-light photoreceptor is responsible for the low and the high light intensity responses (Haupt 1965).

In *Boergesenia* blue light inhibits the growth of the rhizoids. The negative phototropism in these organs results from the creation of a new growing point on the flank farthest from the light source (Ishizawa & Wada 1979).

Mosses and liverworts

In mosses and liverworts the gametophyte plant develops from an initial filamentous stage called the protonema, and these protonemal filaments generally show strongly positive bulging responses to light. The response is induced by different types of light in different organisms, e.g. by blue light in the liverwort *Sphaerocarpus* (Steiner 1969) and the moss *Funaria* (Jaffe & Etzold 1965), but by red light (a phytochrome-mediated response) in the moss *Physcomitrium* (Nebel 1968). In liverworts the subsequent development of dorsiventrality in the thallus is also governed by light, and the thallus itself is diaphototropic; the rhizoids on the shaded side of the thallus are negatively phototropic (Banbury 1959). In mosses, the eventual gametophyte plant is also strongly phototropic, although this is often masked by a dense, tufted growth habit. And in both liverworts and mosses, the seta or stalk of the sporophyte is positively phototropic (Banbury 1959).

Ferns

In ferns, light is involved in several aspects of development and orienta-tion (Miller 1968). After spore germination, if the germ tubes are maintained under red light, only long protonemal filaments are pro-duced (Fig. 4.1a). These protonemata are strongly phototropic by bulging responses (Fig. 4.1b), to red light in *Pteridium* (Davis 1975) and *Adiantum* (Kadota *et al.* 1982) but to red or blue light in *Dryopteris* (Etzold 1965). The *Dryopteris* responses are mediated by two separate photo-receptors, phytochrome and a blue-light receptor. Blue light is also effective in transforming the one-dimensional growth of the filaments into a two-dimensional growth habit to produce the gametophyte thallus (Fig. 4.1c). The rhizoids are negatively phototropic, again in *Dryopteris* by a red-light-induced change in position of the growing point (Hartmann *et al.* 1965). In the sporophyte generation in ferns in general, the rachis (stem) and the pinnae of the fronds show positive phototropic behaviour that becomes stronger with greater development (Banbury 1959).

In the clubmoss *Selaginella*, different types of phototropic responses are obtained depending upon which side of the stem is irradiated (Bilderback 1984). If the dorsal (upper) side of the normally horizontal stem is

(a)

(b)

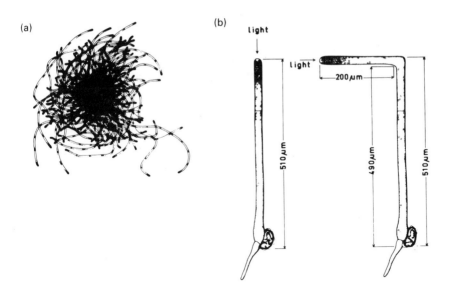

(c)

Figure 4.1 Effects of light on the growth and development of protonemal filaments in the fern *Dryopteris filix-mas*. (a) Filaments grown for 2.5 months under red light. (b) A single filament; 5 days after germination the direction of the light source was changed by 90° (arrows) and growth continued for a further 2 days. (c) Prothallus grown for 2.5 months under blue light. (From Mohr 1972.)

irradiated, there is a slight negative curvature downwards away from the light; if the ventral side is irradiated there is a strong positive curvature down towards the light. Such a differential phototropic response seems to serve the adaptive function of maintaining the dorsal leaves in an orientation in which they are exposed to the maximum amount of light. Both types of response are induced by blue light, and the strong positive curvature requires the presence of the dorsal leaves and may be mediated through the action of auxin (Bilderback, 1984).

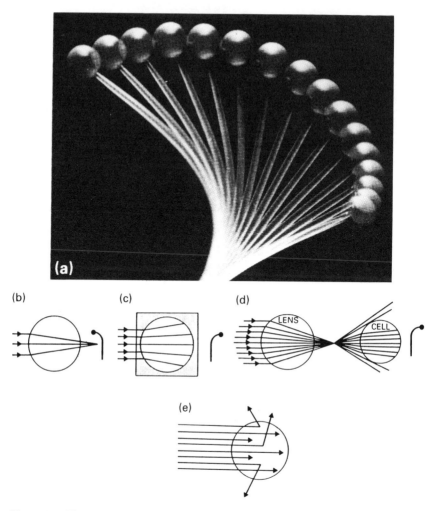

Figure 4.2 Phototropism in the sporangiophore of *Phycomyces*. (a) Phototropic curvature; the source of blue light is on the left, photos were taken at two-minute intervals (from Dennison 1979). (b)–(e) Light reception by the sporangiophore under different conditions: (b) in air, (c) submerged in a medium of greater refractive index, (d) after initial passage of the light through another focusing lens, (e) after absorption of the light by the cell contents (from Shropshire 1962).

Fungi

Directional growth responses to light are found in all groups of (terrestrial) fungi. These responses are usually positively phototropic and their occurrence is largely restricted to reproductive structures, although exceptions to each of these generalizations may exist in the case of leaf pathogens, particularly in the rust fungi, where the vegetative germ tubes of some species may be negatively phototropic (Carlile 1965, Page 1968). The types of reproductive structures showing phototropic responses include unicellular sporangiophores of phycomycetes, and multihyphal structures such as perithecial beaks, asocarps, and stipes in higher fungi. Bowing is thus the most common form of response, even in unicellular structures, although bulging may occur in some immature stages (Page 1968). Another distinguishing feature of phototropic responses in fungi is that they commonly involve lens effects instead of, or in addition to, screening effects. (This means that the light gradient is reversed across the organ, with more light being received by the 'shaded' side.) Phototropic responses in fungi are traditionally considered to be induced by blue light. However, this generalization also may require some qualification. Early studies indicated that the condiophore of *Entomophthora coronata* was responsive towards both blue and red light, behaviour possibly indicating the involvement of a porphyrin-type photoreceptor (Page 1968). More recent studies of such a traditionally 'blue-absorbing' organ as the sporangiophore of *Phycomyces* have indicated the existence of phototropic action maxima at several other wavelengths outside the blue region of the spectrum (discussed in Section 4.2.1).

The sporangiophore of *Phycomyces* is very strongly phototropic, and its responses have been the subject of particularly intensive investigation. In fact, the phototropic response of this organ represents one of the most sensitive biological photoresponses, being activated by a 10 second light pulse which delivers a dose of only 10^{-7} J m^{-2} (this is over 1000 times less than that needed to induce phototropism in the coleoptile). This sporangiophore is a giant unicell which can reach a length of over 10 cm at a growth rate of 3 mm h^{-1}. Its different stages of development are numbered from stage I, in which no sporangium is present, through to stage IV, which carries the sporangium and spores (Fig. 4.2a). In stage IV, growth occurs in a zone 0.2–2.5 mm behind the apex and is accompanied by a clockwise rotation of the organ, fastest in the growing zone at around 10° min^{-1}. The growth rate of the sporangiophore is unaffected by continuous overhead light. However, exposure of the organ to light or even to an increase in light intensity, brings about a transient increase in growth rate known as the 'light growth reaction' (LGR). The response to only a change in light intensity and the transient nature of the response, indicate that the phenomenon of sensory adaptation is involved, although its basis is not clear. The biochemical events of the LGR itself are also not clear, but it is generally accepted that the LGR is responsible for the organ's phototropic behaviour (Dennison 1979, Pohl & Russo 1984). In unilateral light, after a lag of 2–10 min, the organ bends towards the light at about 7° min^{-1}. This phototropic bowing response, in which the side of the organ farthest from the light is growing at a faster rate, is

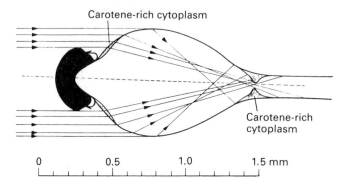

Carotene-rich cytoplasm

Carotene-rich
cytoplasm

0   0.5   1.0   1.5 mm

Figure 4.3 Light reception by the sporangiophore of *Pilobolus*; the sub-sporangial swelling focuses light onto the sensitive growing region; some species also have a 'collar' of a carotene-rich zone (from Banbury 1959).

considered to result from the lens effect of the sporangiophore itself. That is, more light is focused onto the far wall of the organ (Fig. 4.2b) and it experiences a greater LGR. This has been concluded from observations that the phototropic behaviour of the sporangiophore is reversed in situations which tend to negate its own lens effect; these situations include immersion of the sporangiophore in a substance which has a greater refractive index than the organ itself (Fig. 4.2c); unilateral irradiation with light which is first passed through an external lens (Fig. 4.2d); and unilateral irradiation with ultraviolet (UV) light which is strongly absorbed by the high concentration of gallic acid within the sporangiophore, thus establishing a gradient of decreasing light intensity across the organ (Fig. 4.2e). However, many puzzles still remain, a major one being how a transient growth response can result in a continuous phototropic response. One suggestion is that the spiralling mode of growth of the organ means that localized, light-adapted photoreceptor molecules would be continuously moving through a region of increased light intensity (Dennison 1979). Arguments against this 'rotation theory', however, include the findings that phototropic curvature also involves a decreased growth rate on the lighted side, the zone of maximum rotation is not the zone of greatest phototropic sensitivity, and stage I sporangiophores only rotate at $3°$ min^{-1} yet are just as phototropically sensitive as stage IV (Hertel 1980). [The characteristics of a transient change in an activity, in response to a change in level of stimulus, in a moving (spirally growing) organ, are all intriguingly reminiscent of temporal-sensing mechanisms. Other sensory responses shown by the *Phycomyces* sporangiophore include (Russo 1980): weak gravitropism; a 'barrier avoidance' effect, which may involve a volatile growth regulator; 'rheotropism', or growth upwind, which may involve a response to mechanical deformation (Dennison & Roth 1967); and various other light effects on carotenoid synthesis and reproductive development.]

The phototropic responses of the sporangiophore of *Pilobolus* (Fig. 4.3) have also received considerable attention (Page 1968, Hertel 1980, Pohl & Russo 1984), although in this case there is greater controversy concerning the nature of the LGR (positive or negative), the means of establishing the light gradient (lens or screening effects), and the form of phototropic response (bowing or bulging). Events seem to differ according to both species and stage of development. For example, in *P. kleinii* all stages of sporangiophore development show a positive LGR, and achieve a phototropic bowing response through the action of a lens effect. But in *P. longipes*, stage I sporangiophores show a negative LGR, screening effects are probably involved in the light gradient, and it is not entirely clear whether the growth response derives from bowing or bulging (Hertel 1980); but stage IV organs show a positive LGR and seem to respond through a lens effect. This latter phenomenon is thought to involve the region just below the sporangium (Fig. 4.3). The optical properties of this region may act to focus the light onto a particular part of the growth zone. In some species, there is also a dense concentration of carotenoid pigment in the sub-sporangial swelling, and this is thought to act as a screening pigment to further increase the directional sensitivity of the light-detecting system (Page 1968). (The aiming accuracy of the *Pilobolus* sporangiophore is considerable. If presented with two light sources, it aims exactly between them if they are less than 7° arc apart, but at one or the other if they are more than 7° apart.)

In basidiomycetes, the strength of the phototropic response varies according to species. The stipe of the mushroom (*Agaricus*) is only weakly, if at all, phototropic, but those of wood-rotting fungi, and particularly those of the dung-growing *Coprinus* sp. show strong positively phototropic responses (Banbury 1959). [The phototropic responses of these organs often seem to decrease with maturity, in favour of an increasing gravitropism, but this is thought (Carlile 1965) to be a result of the increased shading of the growing region of the stipe by the pileus, or cap, of the organ.] Phototropism in the basidiomycete stipe seems to result generally from an inhibition of growth on the lighted side of the organ, rather than from any stimulation of growth. However, it is worth emphasizing that although the light-induced bowing-type curvatures of these 'multicellular' fungal structures seem, superficially at least, to be analogous to the phototropic responses of higher plants, there are important differences. In the fungal organs, the light responses are highly localized and are restricted only to the regions actually irradiated; and the different parts of the organs behave independently with no overall co-ordination of response or influence of one region on another (Carlile 1965, Page 1968).

4.1.3 General responses in higher plants

In higher plants phototropic responsiveness is especially prominent in young seedlings, where further development is absolutely dependent upon the aerial organs reaching their future energy supply of light as rapidly as possible. In general, stems and other aerial organs are positively phototropic. However, the differential growth that is responsible for positive phototropic curvature can involve different combinations of growth stimulation or growth inhibition (Section 4.3.2). The actual pattern of differential growth underlying a particular response varies according to the type of organ; its stage of development, including whether it is green or etiolated; and the type of light treatment used to induce the response. The characteristics of the light treatment that are important include not only the actual dose of phototropically active blue light but also the modifying effects of red light (see Section 4.2.2).

Leaves are generally plagiotropic or diaphototropic. However, leaves and petioles in many species also show several other major light-induced responses. These include indirect responses to red light by stems and petioles (so-called shade effects), and turgor responses to blue light by leaf pulvini (which form the basis of photonastic and heliotropic behaviour). All these responses are discussed in Section 4.1.5.

Most roots are non-phototropic, although light often enhances their gravitropic responsiveness (Section 3.1.4). However, some roots, particularly those of the Cruciferae and of young sunflower plants, are negatively phototropic.

4.1.4 Variations in response

Developmentally regulated variations in phototropism
The nature of the phototropic response can vary from that expected for a particular type of organ, although such variation generally involves highly specialized or adapted organs. For example, the tendrils of many climbing plants are negatively phototropic (Darwin 1875a). And the hypocotyl of the parasitic mistletoe (*Viscum album*) is also negatively phototropic (Bell & Coombe 1965). [An early report suggested that the prostrate growth habit shown by many plants under high light intensities, in contrast to the erect form shown in shade conditions, was due to negative phototropism (Langham 1941). However, it was later demonstrated (Palmer 1956) that in these cases the prostrate habit was more likely to be due to some form of positive plagiogravitropism.]

The phototropic response can also change during the life of a plant. In ivy (*Hedera helix*), the stem is negatively phototropic in the shade-loving juvenile stage of development (characterized by lobed leaves), although the leaf petioles are positively phototropic; but in adult plants of 10 years

100

or more (characterized by ovate leaves), the stems also become positively phototropic. An even clearer example of variation in phototropic response according to stage of organ development is seen in the flower stalk of ivy-leaved toadflax (*Cymbalaria muralis*). After fertilization the previously positively phototropic stalk bends away from the light to the extent that the ripening seed pod becomes buried in some dark crevice of the wall on which this plant is usually found (Bell & Coombe 1965). A similar type of behaviour is also seen in the rock-rose (*Helianthemum* sp.).

Exogenously induced variations in phototropism

The major exogenous factor that is responsible for variation in the form of the phototropic response is variation in the form of the light treatment itself. In the first place, variations in the quantity of light, in terms of dosage of phototropically active blue light that is applied as the stimulus, can give different types of response; for example, the phototropic response is positive at low light doses, becomes negative if the dose is increased, then positive again at higher doses and longer exposures. Secondly, variation in the quality of light, and particularly exposure to red light, can markedly change the form and sensitivity of the photo-tropic response to blue light. These aspects are discussed under dose–response relationships in Section 4.2.2.

Exposure to red light itself can result in effects on directional growth. As previously described, filaments in several species of algae, mosses, and ferns show direct phototropic responses to unilateral red light through positive effects on the position of the apical growing point (Section 4.1.2). And, under certain circumstances and in certain tissues, red light can induce a form of phototropism in higher plants; mesocotyls of dark-grown maize seedlings show positive curvature towards low intensities of red light (Iino et al. 1985). This unusual response is presumably related to the extreme light sensitivity of this tissue, and is suggested to result from a very localized form of phytochrome-mediated growth inhibition (Hofman & Schäfer 1987). Red light can also influence the orientation of organs, particularly leaves, through non-phototropic 'shade effects' on the growth of sub-tending stems and petioles (see Section 4.1.5).

An unusual phototropic variation has recently been reported to result from unilateral irradiation of a particular region of an organ. In dark-grown oat seedlings, continuous irradiation of the base of the coleoptile with a microbeam (1.0 mm) of blue light (10 μmol m^{-2} s^{-1}) results in a negative phototropic response, to the extent that the whole organ curves completely backwards to form a loop (Taylor et al. 1988). The basis of this effect is quite obscure.

4.1.5 Other orienting effects of light

Induction of cell polarity

Even in simple, unicellular structures that show no significant morphological differentiation, light, and particularly blue light, plays an important role in inducing cell polarity and bringing about an asymmetric
distribution of cell activities (Haupt 1965, Weisenseel 1979). For example,
in the fertilized eggs of *Fucus* the plane of the first cell division always
occurs at right angles to any light gradient that exists across the cell; the
rhizoid then emerges from the region that receives least light and itself
shows negative phototropism. Zygotes of *Pelvetia* and spores of *Funaria*,
Dryopteris, *Osmunda*, and *Equisetum* also put out their germ tubes from
the region of least light absorption. (However, the region of least light
may be the sub-equatorial zone rather than the rear of the cell, due to
intracellular refraction causing the light rays to by-pass this zone.) In
spores of *Botrytis*, on the other hand, the germ tubes emerge from the
region that receives most light, in this case the rear of the spore, again
due to light refraction within the cell.

In the induction of cell polarity the plasma membrane plays a key role.
Ion pumps are developed and redistributed within the membrane, which
leads to the production of transcellular electric currents and the establishment of ion gradients within the cell, particularly gradients of
calcium. The emergence of the germ tube seems generally to occur from
the region of highest internal calcium concentration, and at the point
of entry of electric current into the cell (Weisenseel 1979, Bentrup
1984).

Polarotropism

In polarotropism an organ grows in a plane perpendicular to the
direction of irradiation, and its orientation within that plane is related to
the angle of the electric vector of plane-polarized light (Dennison 1979,
Pohl & Russo 1984). This phenomenon is seen in the responses of many
unicellular organs and protonemal filaments of algae, mosses, and ferns,
but seems to be much less common in fungal hyphae (Pohl & Russo
1984). It usually involves growth at right angles to the electric vector of
plane-polarized light, although the rhizoids of *Dryopteris* grow parallel to
this vector (Fig. 4.4). The polarotropic response can be observed in
Dryopteris if spores are germinated under vertical irradiation by plane-
polarized red light (Dennison 1979). After 8 days or so the protonemal
filaments are all growing parallel to one another, perpendicular to the
electric vector of the light; if the polarizing filter is turned through a
particular angle, the direction of growth of the filaments also changes by
that angle. Polarotropism is considered to be a special case of phototropism, resulting from the action of a dichroic photoreceptor (Mohr

(a)   (b)

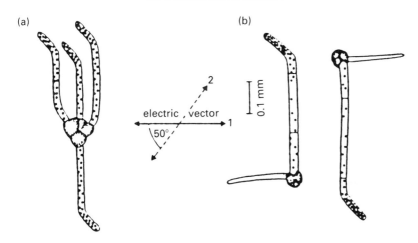

electric , vector

50°

2

1

0.1 mm

Figure 4.4 Polarotropism in protonemal filaments. Filaments were grown under pola-rized light from a direction perpendicular to the page; the change in orientation of the electrical vector of the light is indicated by the arrows. (a) The liverwort *Sphaerocarpus*; (b) the fern *Dryopteris*. (From Pohl & Russo 1984.)

1972), i.e. a receptor whose optical properties and orientation are such that it preferentially absorbs light vibrating in a particular plane.

Many organs that do not show polarotropism nevertheless respond differentially to plane-polarized light, and are said to show 'action dichroism'. For example, if the *Phycomyces* sporangiophore is unilaterally irradiated with plane-polarized blue light, the extent of its phototropic response depends on the orientation of the plane of polarization of the light (Dennison 1979). This type of behaviour is also indicative of the action of a dichroic photoreceptor, and investigation of these responses can provide information about the probable orientations of the photo-receptors within the cell. In particular, such studies have indicated that the dichroic orientation of phytochrome changes during its phototrans-formations (Kadota *et al.* 1982, Haupt 1983).

Photonasty
Photonastic movements of organs occur in response to changes simply in light intensity, rather than light direction. They are shown by the leaves of many species, and are generally based upon turgor changes in the specialized motor cells of the leaf pulvinus. (During leaf opening, so-called extensor cells increase their turgor and volume as a result of potassium influx, while flexor cells contract; during leaf folding, the flexors swell and the extensors contract. In many species, these leaf opening and closing movements follow a circadian rhythm. See Section 1.2.) The pulvinus is not only the region of the photonastic response, but is also the site of stimulus reception, and in many species irradiation of

the pulvinus with blue light brings about leaf opening or closing by inducing the appropriate ion fluxes in the motor cells. Generally, blue light seems to stimulate the influx of potassium into the extensor cells (Koller 1986).

There are many forms of photonastic response. For example, in the shade plant wood sorrel (*Oxalis acetosella*), the individual leaflets fold protectively inwards along the petiole axis in high light intensities. This type of photonastic avoidance response has been particularly studied in *O. oregana* (Briggs & Iino 1983). In this species, the response is initiated after a lag of about 10 sec and occurs fairly rapidly at a rate of around $20°\ h^{-1}$, so that the leaflets have folded within about 6 min.

Some species have further features, either anatomical or physiological, which result in greater differential behaviour across the cells of the pulvinus. This means that greater directional refinement of response is achieved, and a phototropic ability (Koller 1986) can be introduced into the photonastic response. There is thus a gradation, or spectrum, of directional sensitivity between 'open and shut' photonastic responses and 'heliotropic' photonastic responses in which the sensitivity towards light intensity is such that the leaves continually adjust to the changing position of the light source.

Heliotropism

The term heliotropism is used to describe the daily sun-tracking movements shown by many leaves and flowers. In leaves, such movements can fulfil different functions, or confer different types of adaptive advantage, depending on the species and its situation (see Prichard & Forseth 1988). In *diaheliotropism*, the leaf lamina is maintained at right angles to the direction of light. This maximizes light interception and the rate of photosynthesis, and this form of heliotropism is commonly shown by, though not restricted to, plants adapted to shade or to short growing seasons. In *paraheliotropism*, the leaf blade is maintained parallel to the direction of light. The heat load on the leaf is thus reduced and the loss of water minimized. This form of heliotropism can also be found in shade plants, as a means of protection against the effects of sunflecks (Prichard & Forseth 1988). Some species show both forms of heliotropism, e.g. diaheliotropism in the morning and evening, paraheliotropism around midday (see Koller 1986). The means by which this is achieved is not always clear. In some cases, it may be related to the circadian rhythm of the plant. In other cases, it seems more clearly associated with the water status of the plant, e.g. the Australian forage plant, siratro (*Macroptilium atropurpureum*), shows diaheliotropism in well-watered conditions but paraheliotropism in drought conditions (Sherriff & Ludlow 1985). [Similar types of adaptive advantage are achieved by different means in the so-called *compass plants* in which the leaves are maintained in some advantageous fixed orientation (Werk & Ehleringer 1984). For example, in open ground, the leaves of prickly lettuce (*Lactuca serriola*) are aligned with their laminae facing E–W and their edges pointing N–S, thus maximizing light interception in the morning and evening but minimizing the heat load at midday. The means by which this fixed orientation is brought about is unclear, but light seems to play some role, since under shade the leaves are oriented randomly.]

Heliotropic leaf movements occur by several different means, and can involve twisting, photostrophic movements of the stem or petiole (Snow 1962). However, the most extensively investigated examples of heliotropism are those in which the movements result from controlled changes in the turgor of the cells in some sort of pulvinar region. But within this form different mechanisms are involved.

In the Leguminoseae and Oxalidaceae, the leaf pulvinus is the organ of both stimulus reception and physiological response (Koller 1986). The pulvini in these families are generally bilaterally symmetrical, and other features are often overlaid on this to give some degree of control over the upwards and downwards orientation of the leaf. For example, in *Phaseolus multiflorus*, the abaxial surface of the pulvinus is more sensitive to light. In *P. vulgaris*, the adaxial surface of the pulvinus can become shaded by the individual leaflets and the response is therefore modulated according to the combined effects of leaflet position and angle of the light source (Koller 1986). In *Lupinus succulentus*, blue light acts (possibly through some dichroic mechanism) on the individual pulvinules to orient each leaflet independently (Vogelmann 1984). Thus, there seem to be various ways by which a basically photonastic response can be directionally modulated to give some degree of heliotropic control.

In other diaheliotropic species, the leaf lamina itself is the site of stimulus reception. In this respect Haberlandt (1914) observed that in some of these species the epidermis consists of papillate cells (e.g. this is responsible for the 'velvety' appearance of the leaves of *Begonia rex*). He suggested that the domed outer walls of these cells acted as minute lenses, and that leaf orientation was continually adjusted to keep the light rays focused on a particular region of the cell. However, not all sun-tracking leaves in which the lamina is the light-sensitive organ have such protruding epidermal cells.

In *Lavatera cretica*, shading experiments have shown that it is the major veins that are responsible for the detection of the light vector (Koller 1986). In these leaves the palmate venation coalesces in the vascular core of a radially symmetrical pulvinar segment of the petiole. Light-induced differences in turgor in the veins are thus transmitted back to bring about changes in turgor in particular parts of the pulvinus (Koller *et al.* 1985). The initial turgor differentials have been suggested to result from the

Figure 4.5 Photocurvature in cucumber seedlings in non-unilateral light; curvature was induced over 8 h by covering the left-hand cotyledons with aluminium foil (from Shuttleworth & Black 1977).

dichroic actions of photoreceptors oriented in a particular fashion along the veins (Koller 1986). [Another aspect of the *Lavatera* response is that during the night the leaf takes up a position facing the direction of sunrise. However the astonishing feature is that after the plant has been turned through 180°, the leaves still return to ('remember') the original pre-sunrise direction for the following few days.]

The heliotropic movements of flowers are usually due to photostrophic twisting effects, and often involve growth. The adaptive value of such movements is well illustrated by the behaviour of certain Arctic species. In *Dryas* for example, constant alignment to the Sun results in the temperature inside the floral cup being kept 5–10° higher than air temperature. This not only seems to attract more insect pollinators, but also results in more rapid seed development (Ehleringer & Forseth 1980).

Shade effects
The light environment under a vegetation canopy is characterized by its low level of red light relative to other wavelengths, i.e. it has a low R:Fr ratio the actual value of which depends on the density of shading (Smith 1982). This quality of light establishes a low $P_{fr}:P_{total}$ ratio in plant tissues, i.e. a relatively low level of the biologically active form of phytochrome

106

(see Box 4.2). In many species, this situation enhances or initiates various types of shade avoidance responses, some of which involve directional growth.

In the first place, a low $P_{fr}:P_{total}$ ratio seems to enhance the phototropic response to blue light (Woitzik & Mohr 1988a). The effects of red light on various aspects of the phototropic system are discussed further in Section 4.2.2.

However, red light itself, or the lack of it, can also indirectly influence the direction of growth of an organ. This can be demonstrated by artificially darkening one of the pair of cotyledons on a young cucumber seedling (Fig. 4.5); the seedling soon curves in the direction of the lighted cotyledon, due to the greater growth of the stem on the side beneath the shaded cotyledon. This directional response is induced by red light rather than blue light, and is thought to be due to different amounts of growth regulator being supplied from differentially illuminated leaves (Shuttleworth & Black 1977). Again, if one half of a leaf on a *Xanthium* plant is shaded, the sub-tending petiole curves away from that side and that half of the leaf itself curves upwards (Salisbury & Ross 1985).

These various kinds of responses presumably all contribute to shade avoidance patterns of growth in stems and petioles, and to the establishment of leaf mosaic arrangements in canopies.

Box 4.1 Some background to light action and measurement in photobiology

Radiant energy can be characterized by its wavelength, usually given in nm. However, it interacts with matter as whole units of energy called quanta, or photons if they are in the visible range of the spectrum. The energy of a photon is inversely proportional to its wavelength, e.g. photons of blue light have inherently greater energies than photons of red light.

For light to be absorbed by a molecule, and bring about chemical change, the energy of the impinging photon must be exactly equal to the difference in energy level between the ground state of the molecule and a permissible excited state. Therefore any one type of molecule can only absorb certain wavelengths of energy, i.e. it has a characteristic absorption spectrum.

A photobiological response is initiated by a specific type of photoreceptor molecule, and therefore each type of photoresponse has its own characteristic action spectrum which is usually closely related to the absorption spectrum of the photoreceptor (as discussed in the main text, several factors can prevent an exact match.) There are thus two features of light which must be clearly distinguished:

(a) The *energy level* of the photon specifies the type of light, and determines the likelihood of those particular photons being absorbed by a particular type of molecule.

(b)   The *number of photons* specifies the amount of light, and determines the number of absorbing molecules that become photo-excited, i.e. the amount of photochemical change; one photon excites one molecule.

There are many types of photochemical reaction, but two are of particular importance in biology. In photosynthesis, the primary event is a *photo-oxidation*; in vision, the primary event is a *photo-isomerization*. In phototropism, the nature of the primary event is not at all clear but, by analogy with other systems, it is likely to involve a photo-induced configurational change in a membrane-associated receptor, with consequent changes in the properties of the membrane.

   Three different systems of measurement can be used to determine light quantity, but they are not all appropriate for plant biology. Photometric measurements are made with light meters, and measure units of 'brightness' (foot candles, lux, lumens) in relation to the sensitivity of the photoreceptive pigments of the human eye; this system is therefore quite inappropriate for dealing with light in relation to the photoreceptive pigments of plants. Radiometric measurements are made with radiometers, solarimeters, or thermopiles, and measure units of energy (Joules, watts). These measurements are useful in ecophysiological studies where, say, light energy input is to be related to biomass yield. But they should be used with caution in photomorphogenic studies since they take no account of the inherently different energy levels in different types of light (e.g. in treatments with equal energies of red and of blue light there are more photons present in the red light treatment). Quantum measurements are made with a quantum sensor, and measure the actual number of photons. The quantum sensor consists of a filter-corrected photo-cell which emits electrons in response to being struck by photons; the measurable electric current that is generated is therefore directly related to the number of photons per area per time (i.e. amount of light). The unit of measurement is the 'mole of light' (i.e. Avogadro's number of photons, 6×10^{23}). The use of the mole derives from the fact that one photon excites one molecule; thus amount of light can be directly related to the amount of chemical change. (Since the mole is such a large number, the photons present in usual amounts of light are expressed as μmole; in older texts the mole is referred to as an 'einstein', and μmol = μE.)

   Different expressions specify the amount of light in different ways for different purposes:

(a)   *fluence* = quantity per area (μmol m^{-2}) = 'dose';
(b)   *rate* = quantity per time (μmol s^{-1}) = 'flow' (e.g. lamp output);
(c)   *fluence rate* = quantity per area per time (μmol m^{-2} s^{-1}).

In photobiology the 'light intensity' is expressed as the number of photons per area per time, i.e. as the photon fluence rate.

   A more detailed introductory account of all these aspects is given in Hart (1988).

4.2   STIMULUS RECEPTION AND TRANSFORMATION

4.2.1   *Photoreception*

Regions of phototropic sensitivity
It is often popularly asserted that the site of reception of a phototropic stimulus is the tip or apex of an organ. This misconception probably arises from the work of Darwin (1880) who carried out two types of experiments with coleoptiles in this respect. First, he buried seedlings at various depths in fine sand and found that even when only the tip of the coleoptile was exposed to unilateral illumination phototropic curvature still occurred in those regions of the organ that were covered with sand. Conversely, he also demonstrated that when the tips of coleoptiles were covered with small caps of black paper before being exposed to unilateral light, phototropic curvature did not (usually) occur. These results obviously show that the tip of the coleoptile is phototropically sensitive and, when Darwin considered this aspect in relation to his other studies on the involvement of the root tip in the gravitropism of that organ, he concluded that some growth-regulating 'influence' was transmitted from the tip of a tropically stimulated organ. However, although this properly emphasizes the importance of the tip to the overall tropic response, Darwin himself never claimed that it is *only* the tip of the coleoptile that is phototropically sensitive. And, in fact, several subsequent investigations have shown that black-capped coleoptiles can respond phototropically to types of light treatments that are different from those that were normally used by Darwin (see Section 4.3.1).

The apical region is the most phototropically sensitive part of the coleoptile, but the rest of the organ also shows sensitivity to different degrees. It is not actually the extreme tip of the organ that is most sensitive, but a zone about 100 μm behind the tip (Curry 1969). The sensitivity then decreases with increasing distance along the organ, and this decrease is quite dramatic. For example, in the oat coleoptile phototropic sensitivity has fallen by a factor of 36 000 at 2 mm from the apex (Galston 1959).

The situation is less clear in other organs such as stems and petioles. Some early experiments on differential shading or masking of leaves indicated that young dicot seedlings often curved towards a single irradiated leaf or cotyledon (references cited in Pickard 1985b). However, these effects were probably due to red light (Shuttleworth & Black 1977, see also the discussion on 'shading effects' in Section 4.1.5). It is now generally considered that reception of the phototropic light stimulus occurs within the responding stem or hypocotyl itself, but it is not known whether there are regions of different sensitivity within an organ. In this area of investigation, as in analogous investigations of gravitropism, it is experimentally difficult to distinguish between differences in a region's ability to sense the stimulus and differences in its ability to respond to the stimulus. (For example, if, through excision or some other treatment, an organ has lost the capacity for growth, it is then difficult to assess whether it still has the ability to sense a stimulus that influences growth.) However, all the growing regions of an organ seem able to respond to some extent to a directional light stimulus. Masking the apex seems to have little effect on the phototropic response of the stem (Brennan *et al.* 1976). Excision of the apex usually results in a more marked reduction in phototropic curvature, but, as long as some capacity for continued growth remains, never completely abolishes it (Firn & Digby 1980). Therefore the ability to sense a phototropic stimulus is believed to be a general property of all regions of a stem or hypocotyl, though the apex and leaves may supply factors necessary for growth (Pickard 1985b). (The leaves, of course, are responsible for detecting the light vector in many cases of heliotropism, see Section 4.1.5).

The photoreceptor
In photobiological responses, some clues about the nature of the receptor can usually be obtained from knowledge of the particular wavelengths of light that are most effective in inducing the response. Ideally of course, this action spectrum for the response should exactly match the absorption spectrum of the photoreceptor (see Box 4.1). In practice, however, such a match is never achieved. This is due to the involvement of one or more of the following factors:

(a) Lack of knowledge of the precise and appropriate absorption spectrum of the pigment suspected to be the receptor. (The fine details of a pigment's absorption spectrum are greatly influenced by the molecular environment of the pigment; *in vitro* by the type of solvent in which the pigment is dissolved, *in vivo* by possible association with other biological molecules.)
(b) Interference from other pigments. (The presence of other pigments can distort the true shape of the action spectrum by effects such as screening, light scattering, and fluorescence; in green tissues these effects can be particularly significant due to the massive amounts of photosynthetic pigments.)
(c) Involvement of more than one pigment. (If the biological response is the result of the actions of two or more receptors, the action spectrum for the response will not match the absorption spectrum of any of the individual receptors.)
(d) Lack of precision in the measurement of the response. (The fine detail of an action spectrum may be blurred by the fact that the measured response is far removed in terms of intermediary events

110

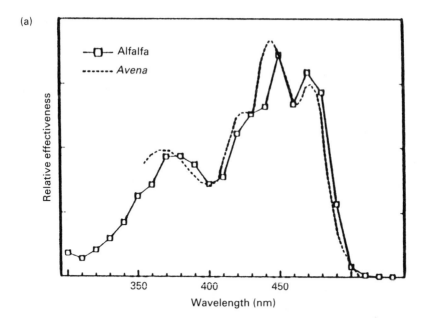

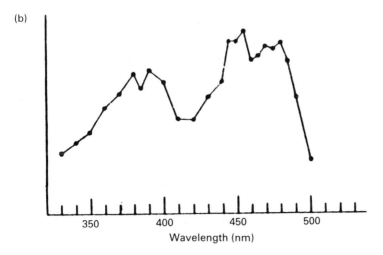

Figure 4.6 Phototropic action spectra. (a) First positive curvature in oat coleoptiles, after Thimann & Curry (————), and in alfalfa (–□–) (from Baskin & Iino 1987). (b) Photocurvature in the *Phycomyces* sporangiophore (from Delbrück & Shropshire 1960).

and processes from the initial response actually induced by the action of the photoreceptor.)

It is therefore difficult to be confident about the significance of the finer details of an action spectrum, or of comparisons between action spectra from different organisms or for different responses. Furthermore, comparisons of action spectra alone only allow conclusions to be made about the possible chromophores of particular photoreceptors, but not about the protein moieties, i.e. quite different types of photoreceptor could utilize the same type of chromophore.

However, with these qualifications in mind, the action spectra for phototropism in coleoptiles (Curry 1969, Dennison 1979) and in a dicot seedling (Baskin & Iino 1987) show the following features (Fig. 4.6a). There is a large peak of activity in the near-UV around λ 370 nm, and three peaks in the blue region consisting of major peaks at λ 450 nm and λ 475 nm and a shoulder at λ 420 nm. Recent studies have shown that, at least in some species, curvature can also be induced by wavelengths up to λ 540 nm in the green part of the spectrum (Steinitz *et al.* 1985). Curvature can also be induced in coleoptiles and in dicot seedlings by unilateral UV irradiation with peak activity around λ 285 nm (Curry 1969) but analyses of the fluence–response kinetics of this form of curvature suggest that it is a separate response that results simply from non-specific damage to the growth process on the UV-irradiated side of the organ (Baskin & Iino 1987). (The action spectrum for phototropism in the sporangiophore of *Phycomyces* is slightly different and is discussed later.)

There are several types of biological molecules that absorb blue light, including pteridines, flavonoids, folic acid, and vitamin K, but on the basis of action spectra, only carotenoid and flavin types of pigments can be considered as candidates for phototropic receptors. And, as with other plant responses to blue light (see Box 4.2), the question of which pigment is actually the photoreceptor is unresolved (Presti 1983, Briggs & Iino 1983).

In the 1930s some form of carotenoid pigment was first considered to be the phototropic receptor, not only because of their absorption of blue light but also because this type of pigment is the photoreceptor in animal vision. In addition, the distribution of carotenes in the coleoptile was found to match the distribution of phototropic sensitivity, with the greatest concentrations of carotene being localized just behind the tip of the organ (Curry 1969). The three peaks of phototropic action in the blue region of the spectrum also provide a good match with the absorption spectrum of carotene (Fig. 4.7a). However, in aqueous solution carotenes do not absorb in the region of the λ 370 nm action peak, and this is considered to be strong evidence against them being the receptor molecules. Carotenes can, in fact, absorb wavelengths in the near-UV

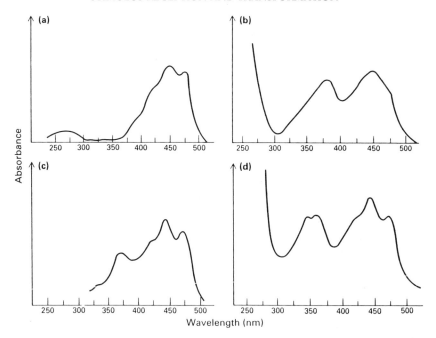

Figure 4.7 Absorption spectra of various preparations of carotenoids and flavins: (a) aqueous solution of β-carotene; (b) aqueous solution of riboflavin; (c) the carotenoid lutein in 54.7% ethanol; (d) riboflavin in ethanol at 77° K (from Hart 1988).

under certain conditions (e.g. Fig. 4.7b), but other evidence against carotene as receptor includes the continued presence of some form of phototropic response in tissues in which the synthesis of carotenoids has been chemically inhibited by treatment with the herbicide Norflurazon (Vierstra & Poff 1981a).

In the 1940s, Galston suggested that a flavin pigment may be the phototropic receptor, on the basis of his observation that, in solution, riboflavin photosensitized the oxidation of IAA. (It is now generally accepted, however, that the photodegradation of auxin is not the cause of the differential growth underlying curvature.) In addition, flavins in aqueous solution absorb strongly in the spectral region of the phototropic action peak in the near-UV around $\lambda\,370$ nm (Fig. 4.7c). And treatment of tissues with chemicals such as potassium iodide, which inhibit photochemical activity of flavins but not of carotenes, leads to a loss of phototropic response (Vierstra & Poff 1981b). However, evidence against a flavin being the photoreceptor is seen in the poor match between flavin absorption and phototropic action spectra in the blue region of the spectrum (Fig. 4.6a v. Fig. 4.7b), although again, under certain conditions, three absorption peaks in this region can be obtained (Fig. 4.7d). Therefore in higher plants, there is evidence both for and against carotenes, and for and against flavins, being involved in reception of the phototropic stimulus. It is unlikely that the situation can be resolved by further refinement of action spectra alone, for the reasons given at the start of this section, and new approaches to the problem are required.

In the sporangiophore of *Phycomyces* the general shape of older action spectra for phototropism and other blue-light responses are qualitatively similar to those of higher plants, but are perhaps rather more typical of flavin-type absorption spectra, with a peak in the near-UV around λ 370 nm and only one major peak at λ 470 nm and a minor shoulder at λ 480 nm in the blue region (Fig. 4.6b). Other evidence for a flavin-type photoreceptor in *Phycomyces* is the continued phototropic sensitivity of caroteneless mutants (Presti 1983). And in an auxotrophic mutant for riboflavin, the substitution of the analogue roseoflavin for riboflavin in the culture media resulted in a phototropic action spectrum that corresponded with the absorption spectrum of roseoflavin rather than riboflavin (Otto *et al.* 1981). However, recent studies not only show that there are additional action peaks in wild-type phototropism at λ 414 nm, λ 491 nm, and λ 650 nm, but also that the shape of the action spectrum is highly dependent on the methods and types of light treatment that are used to determine it (Galland 1983, Galland & Lipson 1985a). These results strongly suggest that there is actually more than one type of receptor involved in phototropism in *Phycomyces*. [It has also been pointed out (Galland & Lipson 1985b) that in the long history of genetic analysis of this organism, no single mutant with a complete absence of photosensitivity has ever been reported, a situation which itself suggests that photosensitivity is not conferred simply by a single type of pigment.] The phototropic system in *Phycomyces* is therefore suspected to involve some sort of array of interacting pigments. And in fact, further recent action spectroscopy studies of phototropism in this organism suggest that antagonistically acting photoreceptors may be involved, one acting positively and one acting to inhibit (Löser & Schäfer 1986).

The notion that there is more than one type of photoreceptor active in phototropism may also be relevant to the situation in higher plants where there is spectral evidence for the involvement of both carotenes and flavins. Furthermore, the previously mentioned studies on the effects of Norflurazon inhibition of carotenoid synthesis and the retention of phototropic sensitivity (Vierstra & Poff 1981a), did not give unequivocal results. In tissues treated with the herbicide, phototropism at λ 380 nm seemed to be unaffected but the response at λ 450 nm was reduced. The authors themselves suggested that these results may indicate that carotenes are involved in phototropism by acting as a screen to increase the light gradient across the tissues. [However, since Norflurazon interferes with the phototropic responses of seedlings even when it is applied just shortly before the phototropic stimulus (Konjevic *et al.* 1987), it seems likely that this chemical also affects phototropism in ways other than through inhibition of carotenoid synthesis.]

The light gradient

The means by which light is detected as a gradient across an organ, and by which the gradient may be intensified, are also important aspects of phototropic light reception. There are several mechanisms that can contribute to the non-uniform reception of light in biological material.

In the first place, if the photoreceptor is located in a dichroic orientation, so that light waves vibrating in a particular plane are preferentially absorbed, then the reception of light within the cell will be non-uniform (Haupt 1983). This type of behaviour is probably significant in the phototropic bulging responses of unicells and protonemal filaments in which the position of the apical growing point is regulated by the direction of light. And dichroism must also be involved in the polarotropic responses described in Section 4.1.5. However, although dichroism may result in the non-uniform reception of light within a cylindrical cell or organ, dichroism by itself is unlikely to result in light being sensed as a gradient across the cell or organ; the distribution of excited photoreceptor will be similar at the front and rear of the irradiated tissue.

An actual light gradient across an organ can be created or intensified by just two types of effect (Poff 1983):

(a) Light refraction (Fig. 4.2b) in which the change in refractive index between an organ and its surroundings can result in a difference in light intensity between the two sides of the organ. This focusing effect is involved in creating differences in light intensity across unicellular organs such as rhizoids, sporangiophores, filaments, etc., but the optical situation is more complex in multicellular organs.
(b) Light attenuation (Fig. 4.2d), which itself can result from two types of process: screening occurs through light absorption by pigments, both by the photoreceptor pigment itself and by other 'inactive' pigments; scattering occurs through reflection (by scattering particles larger than the light wavelengths) and diffraction (by particles smaller than the light wavelengths).

Coleoptiles can respond phototropically to differences in light intensity of as little as 20% between each side of the organ (Pickard 1969). However, actual measurements of light gradients across unilaterally irradiated organs indicate that the differences are usually much greater than this. The ratios of light intensity on each side of a unilaterally irradiated organ vary according to the methods used, but range from 50:1, when measured as light transmission through tissue slices (Parsons *et al.* 1984), to around 5:1, when measured by insertion of optical probes into the irradiated tissues (Vogelmann & Haupt 1985). [An ingenious non-intrusive physiological method of measuring light transmission in plant tissues (Parks & Poff 1985) makes use of the *in vivo* assay of phytochrome transformation along the tissues.] These ratios indicate that there is no light focusing effect across a multicellular organ. Further, it is generally felt that gradients of these measured magnitudes are unlikely to arise solely from screening effects, and therefore that scattering plays a major role in these organs (Parsons *et al.* 1984). In any situation, the amount of scattering is inversely related to the wavelength of the light; that is, scattering effects impose a greater gradient on blue than on red light. However, the involvement of blue light as the major directional light signal in multicellular organs is probably due to more than just this effect.

Red light has very significant effects on growth, and gradients of red light demonstrably exist across the tissues of unilaterally red-irradiated organs (Shropshire & Mohr 1970). However, except under special circumstances (see Section 4.1.4), unilateral irradiation of higher plants with red light does not directly induce phototropic curvature. This suggests that besides the actual light gradient, phototropism also requires what could be termed a 'transduction gradient'; that is, the physiological effects of light must be expressed in a localized fashion, rather than being dispersed across the tissues. The reason for blue light, but not red light, being able to generate this transduction gradient is not clear.

4.2.2 *Type of light treatment and dose–response relationships*

An early requirement in any physiological investigation is an understanding of the quantitative relationship between the extent of the response and the size of the stimulus or factor that brings it about. In higher plant phototropism, however, this relationship between response and light 'dose' is problematical in several interrelated ways.

The first problem concerns the question of dose reciprocity between the intensity and the duration of the phototropic light stimulus. That is, in a straightforward situation a constant level of intensity × duration should give a constant level of response; long exposures at low intensity should give the same level of response as shorter exposures at high intensity, as long as the product of intensity × duration is the same. In the phototropic responses of higher plants, dose reciprocity seems to operate in relation to some types of light treatment, but not in others.

A second problem concerns the question of response to increasing light dose. Again, in a straightforward situation, as dose is increased so response increases, perhaps not linearly but at least in some regular fashion. In phototropism, there is a complex relationship between dose and response in that as dose is increased, the extent of response increases, then decreases, then increases again.

In the early studies of phototropism such complications were not observed, largely because of the simple methods of experimentation that were used, e.g. Darwin merely observed curvature under relatively long-term exposures to continuous light. In the 1900s the Dutch physiologist, A. H. Blaauw initiated a more quantitative approach and, by the use of limited exposures to low levels of light, he related the extent of response to the amount of light received. Further, at light doses *which gave a threshold response* he found reciprocity between light intensity and time, indicating, under those conditions, the involvement of only one photoreceptor. However, subsequent investigations of the dose–response relationship at higher doses revealed the more complex situation of increasing and decreasing levels of response.

The dose–response curve in coleoptiles

The phototropic responses of oat coleoptiles to various light doses are shown in Figure 4.8a. Curve 3 indicates that at a fluence rate of 0.038 W m^{-2} (i.e. 0.14 μmol m^{-2} s^{-1}), when the fluence (i.e. dose) is increased by extending the duration of exposure, the response increases up to a maximum at a fluence of 0.1 J m^{-2} (0.4 μmol m^{-2}); further increase in fluence results in a decrease in response and even curvature away from the light, before the extent of curvature again increases. These different forms of response are termed, respectively: first positive, first negative, and second positive curvatures. Note that at lower fluence rates (Fig. 4.8a, curves 1 and 2), longer exposures are necessary to achieve equivalent fluences, and in these situations the distinction between first and second positive curvatures disappears.

The extent of first positive curvature is proportional to the light dose, and it shows reciprocity between time and fluence. Second positive type curvature appears under 'long' exposures (at a fluence rate of 0.03 W m^{-2} or 0.1 μmol m^{-2} s^{-1}, anything longer than about three minutes), and this form of curvature does not show reciprocity but, instead, is time dependent; the extent of response is determined by the period of exposure. (Although phototropic curvatures under natural conditions must, of course, be due to second positive curvatures, it is actually the first positive type that has received most attention.)

Thus in dose–response studies of phototropism the means of delivering the dose is critical, and the different forms of response are only seen if the light intensity is such that relatively short exposures can deliver appropriate fluences. For example, distinction between first and second positive curvatures is generally carried out by working at some fixed fluence rate and varying the dose by changing the length of exposure. In

(a)

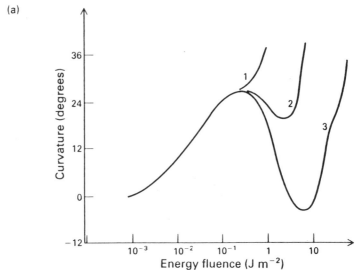

Figure 4.8 Fluence response curves for phototropism in oat coleoptiles. (a) Curves 1, 2, and 3 were determined at, respectively, fluence rates of 3.84×10^{-4}, $\times 10^{-3}$, and $\times 10^{-2}$ W m^{-2} at λ 436 nm (from Dennison 1984, after Zimmerman & Briggs). (b) Response as a function both of fluence rate (increasing from back to front) and of exposure time (increasing from left to right). (From Blaauw & Blaauw-Jansen 1970.)

most organs, the optimal dose for first positive curvature is around 3 μmol m^{-2}, delivered in about 100 seconds (Baskin 1986). Fluence rates which deliver too great a dose for first positive curvature, or fluence rates which require too long a period of exposure to irradiation, will not allow distinction between the different curvatures.

An extremely detailed study of the dose–response relationship was carried out by Blaauw & Blaauw-Jansen (1970). They measured coleoptile curvature as a function of several fluence rates and at many different exposure times. The three-dimensional depiction of their results, termed the 'fluence response surface' (Fig. 4.8b), allows the response to be considered in relation to fluence rate (increasing front to rear in Fig. 4.8b) or in relation to duration of exposure (increasing left to right). At the highest fluence rate (front 10 W m^{-2}), increasing periods of exposure result in the familiar pattern of maximum and minimum responses (maxima being referred to by these authors as regions A, B and C). At lower fluence rates (rear), regions A and B shift to longer exposures (reciprocity holds), but region C does not; thus region C is analogous to the original second positive type of curvature. When the response is considered in relation to the period of exposure, short exposures (left side) give good curvature at high fluence rates, but at longer exposures (right side) high fluence rates give less curvature than low fluence rates. These results are in general agreement with, and extend, the previous findings. They also re-emphasize the importance of light intensity in certain phototropic situations, and the importance of length of light exposure in others.

The reasons for this complexity of dose–response are not at all clear. From their original analysis, Zimmerman & Briggs (1963) suggested that the different types of response may result from the activation of more than one light reaction. And Pickard (1985b) also allows the possibility that there may be two populations of phototropic receptor, or one receptor in different states. However, the action spectra for first and second positive curvature seem similar (Dennison 1979, Pickard 1985b), and a recent study on the induction of phototropism in response to pulsed light treatments also concluded that first and second positive responses shared the same photosystem (Steinitz & Poff 1986).

Therefore, other workers have attempted to account for the dose–response relationship on the basis of a single photosystem. Blaauw & Blaauw-Jansen (1970) suggested that it may be due to differences between light and dark-adapted forms of the photoreceptor. Iino (1987) has developed this idea further and, on the basis of mathematical modelling, has proposed that second positive curvature results from the photosystem undergoing some form of sensory adaptation (see Section 2.1). And Ellis (1987) has suggested that the complexity of the dose–response curve may be related to differences in the levels of excited photoreceptor on opposite sides of the organ, at different light intensities; at high light intensities this differential across the organ may lessen due to photosaturation of the system on the lighted side.

On the other hand, Hofman & Schäfer (1987) have found that red light has different effects on the sensitivity of the ascending part of the phototropic dose–response curve, compared to its effects on the descending part of the curve, and suggest that this indicates the involvement of more than one type of phototropic receptor. These authors also point out that there is good evidence for more than one type of photoreceptor operating in phototropism in *Phycomyces*.

It is also becoming apparent that more than one type of growth response is induced by blue light (see Section 4.3.2), and some of the complexity of the dose–response curve may arise from different growth responses occurring to different extents under different types of light treatment (Firn 1986a).

Effects of red light

Exposure to red light has a very marked effect on the phototropic responses to blue light. However, there has been considerable controversy regarding the exact nature of these effects (reviewed in Pohl & Russo 1984, Pickard 1985b). Briefly, some workers (Curry, Briggs & Chon) found that pre-exposure of coleoptiles to red light decreased the sensitivity of the first positive response (i.e. shifted the position of the peak to higher fluences), and increased the sensitivity of second positive curvature, but had little effect on the overall extent of curvature. Other workers (Blaauw & Blaauw-Jansen) reported that red light had no effects on sensitivity, but that it increased the total bending response. Still others (Elliot & Shen-Miller) found no effects on sensitivity and an inhibitory effect on total bending. Pohl & Russo (1984) concluded that: 'It is clear that red light has some effect on phototropism. However this effect varies from one laboratory to another.'

The results of a recent study, albeit in a dicot organ, may explain some aspects of these past differences (Woitzik & Mohr 1988a). In hypocotyls of sesame (*Sesamum indicum*) red light was found to have different effects on different aspects of phototropism:

(a) Pre-irradiation with red light for an hour or so decreased the sensitivity to subsequent phototropic blue light; the authors suggested that this effect acted at the level of the photoreceptor and was probably analogous to the effect of decreasing the sensitivity of the first positive response in coleoptiles.

(b) Once phototropism was underway, treatment with red light increased the rate and extent of the response; the authors suggested that this effect on a later stage of the phototropic sequence may account for the increased sensitivity of the longer-term, time-dependent second positive response.

A further finding in the sesame studies was that there was a directional component in the effects of red light: if the red irradiation was from overhead or from the side opposite to the subsequent phototropic blue light, phototropism was reduced; if the red irradiation was on the same side as the phototropic stimulation, phototropism was enhanced. This confirms previous findings in maize coleoptiles (Hofman & Schäfer 1987). The basis for these directional effects of red light is not clear, although it has been suggested that they may derive from some localization of the action of red light (Hofman & Schäfer 1987).

Thus, the effects of red light on subsequent blue-light-stimulated phototropism vary according to the timing of the red light treatment and the region of the tissue that is irradiated. (Red light also influences the actual pattern or type of differential growth that is responsible for phototropic curvature, see Section 4.3.2.)

Tip and base responses
Coleoptiles can show different forms of curvature that seem to occur in different regions of the organ and that are referred to as tip and base responses. The 'tip' response describes a type of curvature that is induced by low light fluences (in the region of first positive curvatures) and that is slowly propagated along the coleoptile as far as the mid-point of the organ. The 'base' response refers to a more rapid curvature that occurs at higher fluences and that appears along the whole length of the organ.

It has been suggested, however, that these apparent differences in tip and base curvatures are artefacts induced by uncontrolled effects of red light on the general distribution of growth in the coleoptile (Blaauw & Blaauw-Jansen 1970), e.g. red light enhances growth in the apical region and thus may bring about an apparent tip response. However, Pickard (1985b) maintains that there is a real distinction between tip and base responses, possibly due to different populations of photoreceptors.

Stimulus reception in dicots
The characteristics of phototropic stimulation in dicot seedlings have not been as intensively investigated as in the coleoptile. However, in general, they seem to show the same kind of behaviour, with increasing, then decreasing, then again increasing extents of curvature in response to increasing light dose (Pickard 1985b). In most cases the fluence level for first positive curvature is similar to that in coleoptiles (Baskin 1986, Baskin & Iino 1987), i.e. 3 μmol m^{-2} in 100 seconds, with poor or no response at less than 0.03 μmol m^{-2} or more than 300 μmol m^{-2}. In buckwheat (*Fagopyrum esculentum*) the region of first positive curvature seems to extend over a wider range of fluence (Ellis 1984).

There seems to be some variability between species with regard to the

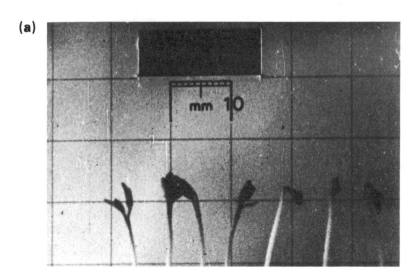

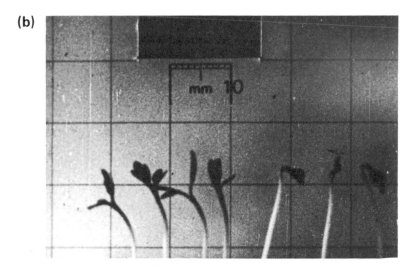

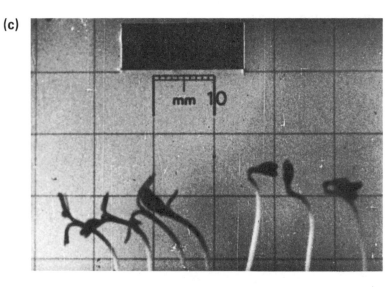

general level of dicot photoresponsiveness. Buckwheat shows a relatively rapid response, beginning curvature after a lag period of about 4 minutes and bending through 90° in less than an hour (Ellis 1984). Many species only begin to show curvature after 60 minutes or so, and sunflower in particular generally takes several hours to develop curvature. (The considerable amplitude of circumnutation that is usually present in sunflower also contributes to the general unsuitability of this species for tropic investigations.)

The phototropic responsiveness of dicot seedlings also varies according to whether they are green or etiolated. Until recently, this aspect has been somewhat controversial, with etiolated seedlings showing greater responsiveness in some species (Everett 1974) but green seedlings being more responsive in others (Hart & MacDonald 1981, Cosgrove 1985). However, it now seems that such differences may have been due to differences in the phototropic light treatments that were used, rather than to differences between species. For example, in buckwheat (Ellis 1987), etiolated seedlings were more responsive than green seedlings to low levels of unilateral irradiation (0.02 μmol m^{-2} s^{-1}), but green seedlings were more responsive than etiolated at high levels of unilateral irradiation or long exposure times (8 μmol m^{-2} s^{-1} or 45 minutes).

The basis of this 'greening' difference in phototropic responsiveness in dicots is not clear. It may be due to effects of de-etiolation on some aspect of the receptor system; these could be direct effects on photoreceptor sensitivity, as envisaged for some of the effects of red light (Woitzik & Mohr 1988a); or they could be indirect effects, such as greater amounts of masking pigments in green seedlings (Ellis 1987). Alternatively, it could be due to differences in the general patterns of growth between etiolated and green seedlings. In this respect, the forms of phototropic curvature shown by etiolated and green seedlings are quite distinct, at least in some species (Fig. 4.9), e.g. in etiolated cress seedlings under continuous unilateral irradiation, curvature starts slowly at the tip and then progresses down the hypocotyl, but in green seedlings curvature appears more rapidly along the whole length of the hypocotyl. (These different forms of curvature seem analogous, superficially at least, to the tip and base responses of coleoptiles.)

4.2.3 Physiological mediation of the stimulus

Little is really known as yet about the means by which the phototropic light stimulus is transformed into a regulatory metabolic signal. It is

Figure 4.9 Time-lapse sequence of phototropic curvature in hypocotyls of etiolated (three plants on the right) and green (four plants on the left) cress seedlings. Plants were exposed to 3 μmol m^{-2} s^{-1} continuous, unilateral blue light. (a) 0 time (darkness), (b) 35 min light exposure, (c) 80 min light exposure. (From Hart 1988.)

possible however, to demonstrate a considerable separation in time between the events of stimulus reception and the growth response, which may prove useful in investigations of this area. If coleoptiles are exposed to unilateral light at 2 °C for 45 minutes, phototropic curvatures can be subsequently expressed in darkness at 20 °C (Pickard 1969); and in fact, after such low-temperature phototropic stimulation, the tissues can be stored in darkness at 2 °C for at least 7 hours and still retain the ability to develop subsequent curvature in darkness at 20 °C.

Within this stage there must also be considerable amplification of the signal. Galston (1959) made some calculations about the probable extent of this amplification and, although they carry many assumptions, they still serve as a useful illustration of the situation. He considered the light dosages of first positive curvature and, with some assumptions about absorption rate, estimated that 2×10^8 quanta of light were absorbed by the receptors. He also considered the extent of differential growth that was induced in the responding organ and calculated, again with a few assumptions, that around 10^{12} molecules of growth regulators must be involved. Thus, 2×10^8 quanta elicit a response that involves at least 10^{12} biomolecules; that is, there has already been a 4000-fold amplification between the reception of the signal and an early physiological stage of the response.

The amplification mechanisms in biological systems include enzymes, growth-regulating molecules (hormones), and changes in the properties of cell membranes. There is no evidence for any unique or special enzyme system being involved in the phototropic response. Of course, hormones in general, and auxin in particular, are associated traditionally with phototropism, and this aspect is discussed in the next section. However, much recent investigation of the effects of blue light on growth and development has centred on light-induced changes in the cell membrane.

The general pivotal role of the cell membrane in the mediation of light stimuli has recently been reviewed (Blatt 1987). In many systems there is evidence that the photoreceptor is located on the membrane, and that light changes the permeability of the membrane towards ions, this change being manifest as a change in the electrical potential of the cell. [A change in the transverse electric potential of the oat coleoptile in response to unilateral irradiation was recorded by Schrank in 1946 (cited in Pohl & Russo 1984), but, with a lag period of 30 minutes, this response, like the original geo-electric effect, was considered to be a result of tropic movement rather than a cause.] In the alga *Vaucheria*, in which blue light affects the position of the apical growing point, it also brings about a change in the inward flow of current at the apex; this blue-light-induced change in current occurs before there is any observable change in growth (Kataoka & Weisenseel 1988). In a perhaps analogous phenomenon, blue light also stimulates the influx of potassium ions into the guard cells of stomata (Zeiger *et al.* 1985). And the red-light-induced movements of the chloroplast in cells of *Mougeotia* seem to involve phytochrome-mediated changes in membrane permeability towards calcium (Blatt 1987). Higher concentrations of (apoplastic) calcium have also been observed on the lighted side of phototropically stimulated sunflower hypocotyls (Goswami & Audus 1976), although these authors themselves doubted the tropic relevance of this finding.

A separate area of investigation at the biochemical level also implicates membrane changes in light responses. Blue light induces changes in the spectrophotometric properties of tissues and tissue extracts, so-called LIACs or 'light induced absorbance changes' (see Senger & Briggs 1981, Schmidt 1983, Pohl & Russo 1984). These changes seem to involve the reduction of a *b*-type cytochrome, and the working hypothesis in these studies is that a flavin photoreceptor, covalently bound to its apoprotein in the cell membrane, is responsible for the photoreduction. [However, Pohl & Russo (1984) caution that such LIACs are also found in the slime mould *Dictystelium*, for which no blue light responses are known, and in human HeLa cells.)

Therefore, several lines of evidence implicate membrane changes in responses to blue light. But there is still a large gap in knowledge between possible events at this level and at the level of phototropic growth responses.

Box 4.2   Light and plant growth

Light is utilized by plants in two general ways, as a source of energy and as a medium through which to receive information about their environment. Their responses to this information are expressed through the processes of photomorphogenesis, i.e. through the control of plant form by non-photosynthetic effects of light on growth and development.

The light environment carries a high content of potential information. The several parameters of light, such as its quality, quantity, direction, and periodicity, can each be modified by other aspects of the environment so that individual habitats have specific and characteristic light regimes, both at particular times of the day and at particular times of the year (Smith 1982, Hart 1988). Plants have the ability to detect and to respond to variations in all of these parameters of the light environment, through the actions of their photoreceptors.

There are two distinct parts to a plant photoreceptor. The chromophore is the pigment part that absorbs particular wavelengths of light and becomes photochemically excited. The apoprotein is the part that is responsible for locating the photoreceptor within the cell, say on a membrane, and that is also involved in transforming the excitation energy of the chromophore into some kind of metabolic energy.

Plants have several different types of photoreceptors. The photosynthetic receptors are responsible for transforming the energy of the light environment into chemical energy, and are localized exclusively in the chloroplasts. The photomorphogenic receptors are responsible for monitoring and processing the information in the light environment, i.e. the variations in light quality, quantity, direction, and periodicity; they include two general classes of receptor, phytochrome and blue-light receptors.

Phytochrome is a bluish chromoprotein that is found in all classes of green plants, including algae, mosses, and ferns; it has not been detected in fungi. The chromophore exists in two isomeric forms that are interconvertible by red and far-red irradiation. The interconversion only occurs if the pigment is attached to its apoprotein, and is usually represented as:

$$P_r \xrightarrow[\lambda \, 730 \text{ nm}]{\lambda \, 660 \text{ nm}} P_{fr}$$

(Both isomers also show significant absorption of blue light.) The far-red absorbing form (P_{fr}) is generally held to be the biologically active form.

The traditional criteria for the identification of a phytochrome-controlled response is the demonstration of reversibility of the response by red and far-red irradiation; this photoreversible mode of phytochrome action is referred to as the 'low-energy response' (LER). However, phytochrome does not simply function in this on–off mode. The absorption spectra of P_r and P_{fr} overlap. Therefore in the natural environment, a particular photoequilibrium or ratio between the two isomeric forms is created according to the exact composition of the light. The actual position of equilibrium (i.e. the relative amount of active P_{fr}) determines the extent of phytochrome action in that situation. Thus, phytochrome provides plants with a means of monitoring and responding to the quality of light in an environment, in terms of the different amounts of red:far-red radiation present, say, in shade conditions or at different times of the day. In this mode of action, sometimes referred to as the 'high-irradiance reaction' (HIR), phytochrome is involved in the regulation of over 100 responses in such areas as germination, stem growth, leaf development, pigment synthesis, and flowering. (Besides the LER and the HIR, there may be a third mode of operation of phytochrome. Growth responses in the mesocotyls of grass seedlings are induced by extremely low levels of red light, around 10^{-4} μmol m^{-2} s^{-1}. This has been termed the 'very low fluence response'. See Mandoli & Briggs 1982, Iino et al. 1984.)

Blue light is involved in a very wide variety of responses, including effects on stem and leaf orientation, stem growth, leaf development, pigment synthesis, and cytoplasmic streaming in higher plants, and effects on cell polarity, reproductive development, sporulation, and pigment synthesis in lower plants and fungi. It is thought highly likely that, in fact, more than one type of photoreceptor is involved in this area, and nowadays the general term 'blue-light receptors' is used rather than the old name of cryptochrome.

Action spectra for the blue-light responses implicate two types of pigments, carotenes and flavins, as possible photoreceptors, and arguments for and against each of these candidates are given in the main text in relation to phototropism (Section 4.2.1). Present opinion tends to favour flavin, but there may well be different types of chromophore in different responses or even interaction between different pigments in a single response.

Treatment with red or with blue light results in the inhibition of growth in stems and hypocotyls, but there are significant differences between the two types of responses (Cosgrove 1985, Hart 1988). Inhibition of growth by red light often takes several hours to be induced, and the inhibitory effect seems capable of being transmitted to non-irradiated tissues. On the other hand, inhibition by blue light is established very rapidly, often within seconds, usually affects only the tissues actually being irradiated, and often disappears soon after the cessation of the light treatment; blue light is also, of course, capable of establishing across and along an organ the differential growth that leads to tropic curvature. The actual mechanism of action of blue light is not yet known, though several lines of evidence implicate effects on the cell membrane (see Section 4.2.3).

4.3 REGULATION OF THE GROWTH RESPONSES IN HIGHER PLANTS

4.3.1 Hormonal involvement

Auxin

In phototropism, as in gravitropism, the traditional Cholodny–Went explanation for the regulation of differential growth during curvature involves:

(a) the development of a lateral auxin asymmetry in the tip;
(b) the longitudinal transmission of this asymmetry;
(c) the consequent development of a trans-organ growth differential.

These ideas, of course, originated from Darwin's observations on the phototropic responses of partially buried and tip-covered coleoptiles. They were developed further in the experiments of Boysen-Jensen and of Páal in the 1920s and culminated in the bioassay demonstrations of Went that there was more growth-promoting activity on the shaded side of phototropically curved coleoptiles (see Section 2.2).

The asymmetry in auxin distribution is considered to result from lateral movement of auxin, rather than from effects on auxin synthesis or degradation, because the total amounts of auxin remain similar before and after phototropic stimulation (Briggs et al. 1957, Briggs 1963). And the tip of the coleoptile is considered to be the region in which lateral movement of auxin occurs, from the results of certain diffusion experi-

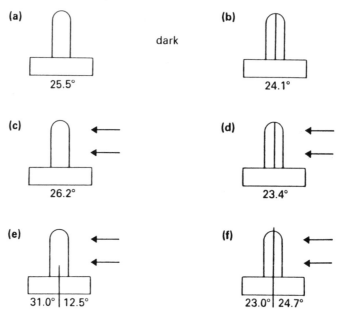

Figure 4.10 Auxin diffusion into agar blocks from variously treated coleoptile tips of maize. The number under each block indicates the degrees of curvature produced by that block in the Went curvature test for auxin; the vertical lines through some of the tips represent impermeable barriers; the arrows indicate the direction of light. (Twice as much auxin was obtained in (e) and (f) because each block had been in contact with six half-tips, the equivalent of the three whole tips of (a)–(d). (From Briggs *et al.* 1957.)

ments (Fig. 4.10). When coleoptile tips were placed on agar 'receiver' blocks and exposed to unilateral light, different amounts of auxin were found to have diffused out of the lighted and shaded sides of the tips; this auxin differential in the receiver blocks was not obtained, however, if an impermeable barrier was longitudinally inserted into each tip. In addition to this effect of unilateral light on the lateral transport of auxin in the tip of a coleoptile, some more recent work suggests that the longitudinal transport of auxin along the irradiated side of an organ may also be inhibited (see Dennison 1984), thus tending to enhance the asymmetry of auxin concentration across the organ. Possible models for the development of an asymmetry in auxin concentration are discussed by Pickard (1985b).

Asymmetries in auxin concentration after phototropic stimulation have also been found within the tissues themselves in a number of investigations. These cases include not only determinations of endogenous auxin content, but also demonstrations of asymmetries in exogenously applied radiolabelled IAA (reviewed in Pickard 1985b). Generally, such auxin asymmetries consist of ratios in auxin contents of the order of 35:65 between the irradiated and shaded sides of the organ. And further circumstantial evidence for the causal involvement of an auxin gradient in tropisms is seen in recent studies which indicate that when auxin is deliberately applied to coleoptiles as a gradient, growth curvature develops basipetally along the organ in a manner analogous to tropic curvature (Baskin *et al.* 1985, 1986).

Nevertheless, as in gravitropism, there are several difficulties with this traditional view:

(a) It has often been demonstrated that decapitated or tip-covered coleoptiles can develop phototropic curvature (Franssen *et al.* 1982, Firn 1986a). These responses are generally of the second positive type, towards continuous unilateral irradiation. They may develop from an auxin gradient created in some other part of the organ (Pickard 1985b), but even so, their occurrence relegates the tip of the organ from its central role in the development of a gradient. [Furthermore, although some early studies reported an auxin asymmetry in organs undergoing second positive curvature (cited in Pickard 1985b), it has been pointed out (Firn 1986a) that the behaviour of auxin during this form of response has not been very intensively studied.]
(b) The causal involvement of a lateral auxin gradient has itself been questioned. Phototropic curvature occurred in the presence of exogenous IAA, at concentrations that were presumed to be swamping any endogenous gradient of auxin (Ball 1962). Alternatively, phototropic curvatures in coleoptiles (Gardner *et al.* 1974) and in hypocotyls (Franssen & Bruinsma 1981) have been observed in the absence of any detectable gradients of IAA.

Two quotations from recent reviews of phototropism perhaps epitomize present opinions about the phototropic auxin gradient:

'It seems that the role of IAA in the phototropism of *Avena* and *Zea* coleoptiles is still an open question.' (Pohl & Russo 1984.)

'. . . it has become clear that the orientation of growth with respect to light, or phototropism, is indeed brought about by a gradient in hormone concentration.' (Pickard 1985b.)

The difference between these statements is not really based upon any new information, but simply reflects the difference of opinion regarding the present evidence.

However, what has become clear in recent years is that phototropic curvature can result from a wide variety of patterns of differential growth (see Section 4.3.2), deriving not so much from differences between species but from differences between experimental protocols and light treatments. These different forms of growth response may be based upon different regulatory mechanisms. Further investigation of the role of auxin in phototropism should therefore be conducted with a greater awareness of possible differences in the type of growth underlying curvature; and greater consideration should be given to the possible regulatory role of the epidermis in phototropism as well as in gravitropism (see Ch. 3), particularly with regard to the effects of light on its properties or auxin content.

Other hormones

Other growth regulators do not seem to play primary roles in phototropism. In phototropically stimulated hypocotyls of sunflower, gradients of gibberellin-like activity have been observed (Phillips 1972b), but unilateral application of gibberellin does not result in tropic-like curvature (Pickard 1985b). Ethylene, as in gravitropism, slows down phototropic curvature, but this seems to be through a general effect on growth rather than being specific to phototropism. No gradients of ABA have been observed in phototropically stimulated organs, but a chemically related inhibitor, xanthoxin, a carotenoid derivative, has been observed to accumulate in greater concentrations on the lighted side of sunflower hypocotyls (Franssen & Bruinsma 1981). However, a general role for this growth inhibitor in phototropism is doubted (Firn 1986a) since plants that were devoid of carotenoids (either by mutation or by chemical inhibition) still showed phototropic responses. The light-induced synthesis of another growth inhibitor, raphanusin, is considered by Japanese workers to be solely responsible for phototropism in radish hypocotyls (Hasegawa et al. 1987). However, the general occurrence of such inhibitory materials is not established, and most workers find that, far from being a phenomenon solely of growth inhibition, phototropism generally involves considerable stimulation of growth.

4.3.2   Growth responses

Phototropic growth

Growth responses during phototropic curvature were first investigated in the early years of this century by A. H. Blaauw, who found strong growth inhibition on the lighted sides of unilaterally illuminated coleoptiles. Further studies were carried out in 1928 by Went and a year later by DuBuy & Neurnbergk (see Went & Thimann 1937), who all found growth inhibition on the lighted side in the apical regions of the coleoptile but growth stimulation on the shaded side in the basal regions. Virtually all subsequent work on phototropism then became interpreted for many years wholly on the basis of growth stimulation on the shaded side of an organ.

In the 1970s, a group of investigators at York University, England, began using modern methods of time-lapse photography to re-investigate the growth events involved in phototropic curvature. They found varying amounts of growth stimulation on the shaded sides of unilaterally illuminated organs, but considered that the major and most consistent form of response was strong growth inhibition on the lighted sides (Franssen et al. 1981, 1982). Some Japanese workers also consider that, in hypocotyls of radish at least, phototropic curvature arises solely from inhibition of growth on the irradiated side of the organ (Hasegawa et al. 1987). However, several other investigators have found that, both in dicot organs and in coleoptiles, phototropism generally involves growth stimulation on the shaded side as well as inhibition on the lighted side (Hart et al. 1982, Iino & Briggs 1984, Baskin et al. 1985, Rich et al. 1985, Baskin 1986).

There is thus a certain amount of inconsistency regarding the actual nature of the growth events responsible for phototropic curvature. Some of this may be due to the organs being at different stages of circumnutatory movement when the phototropic stimulus is initiated (Baskin 1986). However, there are still substantial differences in the form of the overall growth responses, not only between the results of different investigators but also between different types of phototropic light treatments (Hart et al. 1982, MacLeod et al. 1986). It now seems that this is due to the fact that blue light can induce more than one type of growth response.

Blue light and growth

Blaauw investigated the effects of omnilateral blue light on the growth of both the *Phycomyces* sporangiophore and the oat coleoptile (see Went & Thimann 1937). In both organs he observed a response that he called a 'light growth reaction'. In the sporangiophore this was seen as a transient increase in the rate of growth, whereas in the coleoptile it consisted of a temporary decrease in growth followed by a transient stimulation. Blaauw therefore reasoned that, under unilateral light, differential light reception between the two sides of the organ would result in a differential growth rate across the organ, with, in the case of the coleoptile, the side nearer the light source having the greater reduction in growth rate.

In more recent investigations, rapid growth responses to omnilateral blue light have been detected through the use of some form of linear transducer that provides a continuous record of growth. Several investigators have found that exposure of a stem or hypocotyl to blue light often imposes a local and severe inhibition on growth, usually within minutes (see Cosgrove 1981, Gaba & Black 1983).

It was initially speculated that this rapid, blue-light inhibition of

growth, probably analogous with Blaauw's original light growth reaction, may be responsible for phototropic curvature, particularly since
such curvature generally involves a strong element of growth inhibition.
However, several lines of evidence indicate that the blue-light inhibition
of growth is not the (sole) basis of phototropism:

(a) Phototropism is rarely a response only of growth inhibition; generally there is also substantial growth stimulation, and the stimulation on the shaded side of an organ can even be greater than that
which results when the organ is transferred to complete darkness
(Hart *et al.* 1982).
(b) In some material, e.g. etiolated cucumber seedlings, a strong blue-
light inhibition of growth is present, but only a very weak phototropic response (Cosgrove 1985).
(c) Conversely, in other material, e.g. de-etiolated mustard seedlings,
red-light-grown pea seedlings and sesame seedlings, there are
strong phototropic responses, but only very weak forms of blue-light
growth inhibition (Rich *et al.* 1985, Baskin 1986, Woitzik & Mohr
1988a).
(d) The growth responses that result from switching from bilateral to
unilateral irradiation also argue against Blaauw's notion of a direct
effect of light on phototropic growth: when the light on one side of
the organ is switched off, it is the growth on the remaining
(unchanged) irradiated side that becomes inhibited; this suggests
that it is the co-ordinated effects of the light gradient across the organ
that are important, rather than simply an effect of light on growth
(MacLeod *et al.* 1985).

The obvious conclusion from all these studies taken together is that blue
light must be capable of initiating two distinct types of growth responses
(see also Baskin *et al.* 1985, p. 603). First, it can bring about an inhibition
of growth that is characterized by its rapid imposition and localized effect
only upon tissues that are actually irradiated; this 'blue-light growth
response' (BLGR) may result from a direct effect on some growth
process. Secondly, blue light can induce a phototropic growth response
that involves both growth stimulation and inhibition, and this phototropic effect is transmissible to non-irradiated tissue. [Interestingly, over
50 years ago van Overbeek (1933) suggested that phototropic curvature
in radish seedlings resulted from the actions of two types of light effects,
one of which affected the movement of auxin while the other affected the
sensitivity of the tissues to auxin.]
   In addition to there probably being different actions of blue light, it has
been shown recently that phytochrome can also influence the pattern of

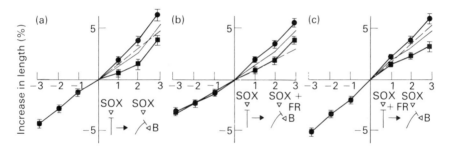

Figure 4.11 Different patterns of growth responses during phototropic curvature of mustard seedlings under different light treatments. At time 0, seedlings were exposed to low fluence rate, unilateral blue light (0.46 μmol m^{-2} s^{-1}) while being maintained under background sodium lighting (SOX), which was or was not mixed with far-red radiation (FR) in order to change the phytochrome photoequilibrium. The graphs show the growth rates of the lit (■) and shaded (●) sides of hypocotyls, the average growth rate (—) and the pre-stimulation growth rate (---). (a) No FR treatment; curvature by stimulation of the shaded side and inhibition of the lit side. (b) Addition of FR; curvature by stimulation of the shaded side. (c) Removal of FR; curvature by inhibition of the lit side. (From Rich *et al.* 1987.)

growth that occurs during blue-light stimulated, phototropic curvature (Fig. 4.11). According to the level of their phytochrome photo-equilibria, phototropic curvature in green mustard seedlings can occur through growth stimulation and inhibition acting together, or largely by stimulation of the shaded side alone, or largely by inhibition of the lighted side (Rich *et al.* 1987). This behaviour may be due to general effects of phytochrome on the overall growth pattern of the hypocotyl, although there may also be more specific effects on the phototropic system, e.g. by affecting the sensitivity of the phototropic receptor or the expression of the phototropic growth responses (see Woitzik & Mohr 1988a, and Section 4.2.3). In any event, these different forms of phototropic growth in response to different irradiation protocols go a long way to providing a possible explanation for the conflicting results of earlier investigations.

Thus in seedling hypocotyls, and perhaps also in coleoptiles, there seem to be three types of growth responses capable of being initiated by light: phytochrome-mediated responses, blue-light growth responses, and phototropic responses. All these responses could also be initiated in certain unilateral light treatments, and the resulting pattern of 'phototropic growth' would depend upon the extent to which each of these types of growth response is expressed. Factors which would influence this include: type of light treatment (including any pre-treatment with red or with white light), type of species (e.g. green seedlings of mustard have no BLGR), and physiological state of the tissue (e.g. etiolated seedlings of many species show a strong BLGR).

A further implication of this situation is that since phototropic curvature in response to blue light can be brought about by different forms of growth response under different types of light treatment, then previous investigations that have considered the phototropic response simply as 'curvature' must be re-examined. (For example, differences in the extent of involvement of the different growth responses may contribute to the complexity of the phototropic dose–response curve. Also, as already pointed out, any considerations of auxin involvement must now also include attention to the details of the irradiation protocol.)

General comments

Light thus controls and influences the orientation of plant organs by several different mechanisms operating at several different levels. In the first place, orientation can be affected by light-induced changes in cell turgor and by changes in growth. Light-induced changes in growth are brought about by different mechanisms in different types of plants, e.g. by a change in the region of growth in protonemal filaments, by asymmetric changes in the rate of growth across fungal hyphae, and by the establishment of differential patterns of growth in the multicellular organs of higher plants. Within the growth responses of higher plants, there are different light-activated mechanisms for establishing differential growth, including the effects of red light on general growth, and the effects of blue light on directional growth. And even within the blue-light induced changes in growth rate, there seem to be different types of growth responses.

   These various effects of light on growth and orientation simply reflect, of course, the crucial role that light plays in the life of plants. However, within this diversity of 'phototropic responses', investigation has so far concentrated largely on a single type of organ (the coleoptile), a single type of response (first positive curvature), and a single type of possible regulatory mechanism (auxin concentration).

4.4   PHOTOTROPISM AND GRAVITROPISM

Most of the research on phototropism and gravitropism has been concerned with the separate investigation of the receptors and responses of the individual systems. However, under natural conditions aerial organs are exposed to simultaneous signals of light and gravity, often acting in different directions, or in different planes. In these sorts of situations, perhaps one of the first questions that comes to mind is simply: which is the more important or more dominant guidance system? Before considering this specific question, it may be useful to briefly review the characteristics of light and gravity with regard both to their general roles in plant growth and development, and their particular roles in acting as sources of directional information.

Light and gravity

Of course, as the ultimate source of energy, light is vital for normal plant growth and development. The light environment, however, shows an extremely wide variation in such parameters as the quality and quantity of light, the periodicity of its presence and absence, and, most significantly in the present context, its direction, which can vary markedly, not only among different habitats but also at different times of the day and at different times of the year.

On the other hand, although gravity obviously influences many aspects of plant form, the successful growth of at least some individuals from seed to fruiting plant in the microgravity environment of a satellite in Earth orbit (Halstead & Dutcher 1987) suggests that gravity is not a vital component for plant development. Moreover, although the magnitude of gravitational force does differ at different points on the Earth's surface, and also shows regular fluctuations throughout the year, such variations are not at all of the same order, or significance, as those of the light environment. For example, gravitational force varies with altitude and with latitude (Tsuboi 1983), but these variations are not known to be of any significance to plant growth. Temporal variations in gravitational force are, of course, manifest in the periodicity of the tides, and such tidal–lunar cycles may be of significance to the temporal behaviour of certain marine algae and diatoms (Cumming & Wagner 1968). However, except for these possible timing effects in a limited number of organisms, gravity seems to interact with plants only as a carrier of directional information.

Therefore, light is vital for plant growth and development but shows great variation in the environment, whereas gravity seems not to be vital for growth and is more or less uniformly and constantly present as a simple directional signal in normal terrestrial habitats. These features must have a considerable bearing on the relationships between the mechanisms that were developed to process each of these forms of directional information.

Phototropism versus gravitropism

Some authors consider gravitropism to be a more universal and fundamental response than phototropism (Jankiewicz 1971, Firn 1986a). However, casual observation of the responses of aerial organs under natural conditions indicates that phototropism generally overrides gravitropism, e.g. a vertical shoot that is continuously exposed to a unilateral light gradient will usually curve away from the vertical, towards the light, to some extent or other. This type of behaviour, in various types of

135

Figure 4.12 Time-lapse sequence of the responses of horizontal, etiolated (without bead markers) and green (with beads) cress seedlings to simultaneous gravi- and photostimuli. Seedlings were placed horizontally and exposed to blue light (10 μmol m^{-2} s^{-1}) from the left. (From Hart & MacDonald 1981.)

organs, has been known for around 100 years (Rothert 1894, cited in Nick & Schäfer 1988).

The actual responses of seedlings that are exposed simultaneously to opposing phototropic and gravitropic stimuli can be observed by time-lapse photography (Hart & MacDonald 1981). In the sequence shown in Figure 4.12, green and etiolated cress seedlings were oriented horizontally and also exposed continuously to unilateral blue light from the side. Initially, in both types of seedling, upward curvature occurs in response to gravity. However, as the sequence continues, the green seedling responds to the light stimulus and reverses its orientation back to the horizontal. The etiolated seedling, in cress at least, does not respond phototropically until after it has been exposed to the light stimulus for several hours (by which time it is no longer etiolated).

At this point some initial, though superficial, answers can be given to the question posed at the beginning of this section. Throughout the plant kingdom, phototropism seems to be a more dominant response than gravitropism. In lower organisms, even in fungi, light seems to provide a more reliable directional signal than gravity towards open air space for spore dispersal, and phototropism generally overrules gravitropism in the orientation of reproductive organs. In seedlings of higher plants, gravity can initiate a tropic response more rapidly than light, but phototropism is generally the dominant response in aerial organs. The extent of this dominance varies with the type of organ and its physiological state. Coleoptiles are altogether more sensitive to the direction of light than they are to gravity. Hypocotyls of many dicot seedlings are initially more gravitropically responsive, but as they become green, so they become phototropically more sensitive and light again assumes its dominant directional role. In mature plants, the situation is much more complex. In lateral branches, orientation is controlled by plagiotropism, which itself is a composite of several types of responses, including nastic growth, gravitropism, phototropism, and shading effects. In leaves and petioles, orientation is more clearly influenced by several light-activated responses, including phototropism, shading effects, and heliotropic effects. The overall orientation of most aerial organs, therefore, depends very much on the particular details of their individual light environment.

The mechanisms that underlie the interactions between phototropism and gravitropism are only just beginning to be investigated. The important factor in the dominance of phototropism is that the light gradient must be maintained, if not continuously, then at least for some considerable period. That is, the response results from the time-integrating action of second positive (type C) phototropic curvature, and actually represents the photogravitropic equilibrium or balance position characteristic of the type of organ, its physiological state, and the details of the lighting conditions.

That it is indeed an equilibrium between phototropism and gravitropism, rather than gravitropism simply being switched off in the presence of light, is apparent from the results of klinostat experiments. On a klinostat, when the antagonistic or 'corrective' influence of gravitropism is minimized, phototropism occurs faster (Pickard 1969) and to a greater extent (Shen-Miller & Gordon 1967, Franssen et al. 1982). This enhancement of phototropism on a klinostat results from effects on both first positive and second positive responses (Heathcote & Bircher 1987). The antagonistic action of gravitropism is most clearly seen when photocurvature is induced by a pulse of light rather than under continuous irradiation. For example, under normal circumstances in maize coleoptiles, a pulse-induced first positive response reaches maximum curvature at about 100 minutes, before gravitropism begins to return the organs to the vertical; but on a klinostat, pulse-induced first positive curvature seems able to be maintained for over 20 hours (Nick & Schäfer 1988).

Analysis of the ways in which first positive curvature and gravitropism interact and influence each other has given some preliminary insight into the general extent of overlap and interaction between the phototropic and gravitropic transduction chains. The initial steps in each process seem to be separate and independent. This was concluded from early investigations that examined the responses to the summation of sub-threshold stimulation, i.e. responses were induced by the summation of the same type of sub-threshold stimulus, but not by the cross-summation of the different types of stimuli (Pekelharing 1909, cited in Nick & Schäfer 1988). This conclusion is supported by the results of a recent study which showed that, if they are applied in opposition, or in parallel at sub-threshold levels, light and gravity act additively and independently on the tropic responses of maize coleoptiles (Nick & Schäfer 1988). However, later steps in each transduction chain seem to be common to phototropism and gravitropism. For example, red light has similar desensitizing effects, both directionally (Woitzik & Mohr 1988a) and kinetically (Nick & Schäfer 1988) on both phototropism and gravitropism, i.e. the site or stage of the red light effect seems to be common to both tropic systems.

Therefore, directional light and gravitational signals may be received and processed by independent systems which then feed into some common physiological system. As long as the light signals are provided continuously, the effects of conflicting gravitational signals are masked, though not entirely absent, and phototropism is dominant.

However, analysis of the interaction between phototropism and gravitropism is at an early stage and we are still a long way from understanding exactly how the orientations of different types of organs are achieved and maintained under natural conditions.

CHAPTER FIVE

Thigmotropism

Thigmotropism is the directional response of a plant organ to touch or physical contact with a solid object, generally through the induction of some pattern of differential growth.

In the popular image, it may seem unusual to consider that plants have a 'sense of touch' (and further, that they 'move' in response to it). But in fact in all plants there is generally some degree of response to contact and other forms of mechanical stimulation, and in some plants this facility has become specialized and developed to confer a particular adaptive advantage. In these special cases, the touch sensitivities of plants compare favourably with those of animals (Shropshire 1979). For example, the human skin can minimally detect a thread weighing 0.002 mg being drawn across it; a feeding tentacle of the insectivorous sundew plant (*Drosera* sp.) responds to a thread of 0.0008 mg, and a climbing tendril of *Sicyos* to one of 0.00025 mg (Darwin 1875a, b).

Because of their relative rapidity, some of the most obvious, and well-studied, plant responses to mechanical stimulation consist of turgor movements (Section 1.2). Although these involve somewhat specialized organs, they may nevertheless be useful in illustrating the general means by which plant cells detect and respond to this form of stimulation, and several of these turgor responses are discussed later in the chapter (Section 5.3.1).

It may also be useful to review briefly the various types of growth responses to mechanical stimulation that are shown by plants in general, before dealing with the details of thigmotropism in particular.

5.1   GENERAL RESPONSES TO MECHANOSTIMULI

5.1.1   *The nature of the stimulus*

There are several different forms of mechanical stimulation that can be experienced by plants. In the past, the prefix 'thigmo-' has been used to denote responses to touch or physical contact, and the prefix 'seismo' for

139

responses to vibration or flexure. Furthermore, some organs show particularly strong responses to contact, whereas others are more responsive to shaking or bending; and usually the different forms of stimulus are not cross-effective on the different types of organ (Pfeffer 1906). However, it now seems that such differences may be more apparent than real, especially in relation to the probable mechanisms of stimulus reception and transduction. That is, both types of stimuli, thigmo- and seismo-, are thought to bring about changes in the cell membrane, contact creating localized pressure differentials and flexure bringing about more generalized effects (Bentrup 1979). Differences in the responsiveness of organs to the different types of stimulus seem to be due simply to different types of morphological or anatomical structures that are particularly suited to focus or amplify a particular form of stimulation (see Section 5.3).

There are also various sources of mechanical stimulation in the plant's environment. Physical contact, either an individual event or some kind of friction, can involve a variety of agencies, including other organisms (e.g. insect pollinators and predators) or inanimate objects (e.g. soil particles). Flexure, or bending, can be slow or rapid, even vibratory, and arises largely from the effects of wind. Pressure can be exerted by external agencies, generally the soil, but it is also a major internal factor within the plant, with compressive and tensile forces being generated by the different rates of growth of the different tissues (see Box 3.2).

Another major source of mechanical stimulation is gravity. In the past, plant responses to gravity have generally been considered to be quite separate from those of thigmo- or seismomorphogenesis. Recently, however, it has been recognized that there is probably a close relationship between the mechanisms involved in the detection of all these types of stimuli (Pickard 1985a, Edwards & Pickard 1987).

5.1.2 Responses in aerial organs

Mechanical stimulation, whether by contact or flexure, usually inhibits extension growth in aerial organs. The effect is described generally as thigmomorphogenesis (Jaffe 1973, 1985), and it can be produced in the laboratory simply by rubbing a growing organ with as few as 5–10 strokes per day. This treatment can inhibit elongation by up to 50%, i.e. it is as effective as light compared to darkness in reducing elongation. The inhibition of elongation is accompanied, both at the cell and at the organ level, by an enhancement of radial expansion. The flexing effects of wind in reducing elongation in the natural environment have also been referred to as seismomorphogenesis (Pappas & Mitchell 1985).

The effects on growth are imposed very rapidly, often within minutes, and are maintained for an hour or so if there is no further stimulation;

subsequent recovery to the original growth rate usually takes at least several hours (Jaffe 1985). Ethylene may be involved in responses to mechanical stimulation (Pickard 1985a, Jaffe 1985), but the rapidity of onset of the growth effects suggests that other factors are also likely to be involved. These may include directly induced changes in the permeability of the cell membrane towards ions or other growth regulators (see Section 5.3).

These effects of mechanical stimulation in producing shorter and thicker organs are probably of great adaptive value in the natural environment, e.g. against the potentially damaging effects of wind. In the laboratory, however, possible effects of unintentional mechanical stimulation during experiments are often overlooked (or ignored). Plants usually need at least several hours to recover from any but the most gentle handling or manipulation (Cosgrove 1981, Gordon et al. 1982).

Besides these general effects on growth, mechanical stimulation can also induce differential growth. In the *Phycomyces* sporangiophore (Dennison & Roth 1967), the dandelion peduncle (Clifford et al. 1982), and the bean stem (Jaffe 1985), lateral flexure of the organ causes growth to be stimulated on the compressed (concave) side and inhibited on the stretched (convex) side, to the extent that the organ subsequently curves towards the side that was stretched. Other directional growth responses to mechanical stimulation include both thigmonastic responses, where the mechanostimulus initiates some endogenously determined pattern of growth, and thigmotropic responses, where the induced pattern of growth bears some relation to the direction of the stimulus (see Section 5.2).

Mechanical stimulation may also exert effects on other aspects of plant development. For example it has been suggested that in Bermuda grass, contact with soil particles is an important factor in determining whether a particular organ develops as a stem or rhizome (Montaldi 1979). And it has been observed that in the woody plant *Liquidamber*, regular but limited shaking seems to induce terminal bud dormancy (Neel & Harris 1971).

5.1.3   Responses in roots

The subterranean environment of most roots must act as a virtually continuous source of mechanostimulation. Darwin (1880) was keenly aware of this aspect, yet in subsequent years the effects of mechanical stimulation on root growth and development have received relatively little attention in comparison to other factors such as oxygen, water supply, nutrition, and so forth. (In fact, there has probably been more investigation of the effects of roots on soil properties, than of soil on root responses.)

141

It is well established that the impedance or resistance offered by the soil has a general inhibitory effect on root elongation (Goss & Russell 1980, Feldman 1984). The effects are seen as decreases in the rates of root extension, and the lengths of individual cells, but increases in cell diameter and cell number, particularly in the cortical tissues. The overall thigmomorphogenic response in roots is therefore similar to that in aerial shoots, i.e. the production of shorter, thicker organs. The extent of the response varies somewhat among species, but in barley, for example, root length can be reduced by as much as 50% as a result of soil impedance (Goss & Russell 1980).

The means by which mechanical impedance exerts its effects on growth are not entirely clear. Indeed, there has been some degree of controversy and confusion regarding the actual levels of impedance that are required to bring about such changes in growth. In an early investigation of the subject, roots were subjected to varying degrees of mechanical stress by being grown in compression chambers filled with glass ballotini beads (Goss & Russell 1980). In these experiments growth became severely limited when very low pressures of 20 kPa were imposed on the chamber and these results have often been quoted as indicating that roots are sensitive to very low levels of mechanical impedance (e.g. Feldman 1984). However, these conclusions have more recently been considered to be incorrect and to be based upon erroneous interpretation of the pressurometer tests, which did not distinguish between the low pressures imposed upon the flexible sides of the chamber and the much higher pressures exerted by the ballotini beads on the roots themselves (Greacen 1986). It is now generally held that roots do not show a drastic response to such low levels of mechanical stress, and that for sustained effects on growth they have to encounter impedances of around 200 kPa (Greacen 1986). Nevertheless, it was noted from the pressure chamber studies that simple physical contact between the root tip and a solid object resulted in a transient inhibition of growth within 10 minutes of contact (Goss & Russell 1980). And the original experiments of Darwin indicated that very small applied pressure differentials across the root tip could bring about marked effects on the direction of growth (see Section 5.2.3). These sorts of response to low levels of mechanical stress suggest that some physiological (hormonal?) mechanism may also be involved. This aspect is considered further in Section 5.3.2.

Another major effect of mechanical impedance seems to be on the induction of lateral roots, resulting both in greater numbers of them and their production in regions nearer the apex of the primary root (Goss 1977), to give an altogether more compact and bushy root system. In this connection, it has been repeatedly observed, both by Sachs and by Noll in the 19th century and by more recent investigators (e.g. Goss & Russell 1980), that lateral roots are induced largely, or even only, on the convex

142

side of a curving main root, whether the curvature is induced by a tropic response or by physical buckling against a barrier; on the other hand, root hairs seem to be produced predominantly on the concave side of the curve. This pattern of development of lateral roots seems to be an effect of the curvature itself, rather than simply a later response to the stimulus that originally produced the curvature (Fortin *et al*. 1989), and may be due to the effects of internally generated mechanical forces on some aspect of hormonal activity.

Therefore in several ways, the full extents of which have probably still to be determined, mechanical factors influence the growth and development of individual roots and play a major role in the overall morphological determination of the root system.

Box 5.1 Some questions about root penetration through the soil

Many studies of root growth through the soil have been largely concerned with measurements of the force that the root may exert against the soil, and of the resistance that the soil may offer to the root. In these latter types of studies, some kind of inanimate penetrometer is often used to measure soil resistance. However, the many changes induced in root growth and development in response to mechanical stimulation perhaps highlight the question of the extent to which force and pressure are actually involved in root penetration through the soil.

There is no doubt that roots are capable of developing considerable pressures, generally of the order of 1000 kPa in a longitudinal direction and 500 kPa radially (see Feldman 1984). On the other hand, transient or directional growth responses may be induced by much smaller externally applied pressures (see Section 5.1.3).

The range of mechano-induced developmental responses that could aid root penetration through soil is fairly wide and includes: changes in root diameter, involving changes in cell shape and cell number; changes in both the extent and pattern of branching; changes in the rate of growth; and changes in the direction of growth (it is not always clear whether the curvature that develops when a root encounters a barrier is a passive buckling response or an active directional growth response). Circumnutation also occurs in roots, but again the extent to which this acts as an aid to soil penetration is undetermined.

Over 100 years ago, Darwin demonstrated the extreme sensitivity and responsiveness of the root. Several years later, Pfeffer initiated penetrometer studies of soil resistance. At the present day there still seems to be these two schools of thought regarding the manner in which roots penetrate the soil: by delicately 'sensing' their way along the line of least resistance, or by forcefully pressuring their way through the resistance. However, roots have to be able to grow through relatively stiff 'structureless' soil (i.e. without visible cracks) as well as structured soils which have some sort of network of continuous pores or cracks. It is therefore not surprising that roots seem to have both the (thigmotropic) ability to make the best use of any available soil structure and the ability to exert considerable growth pressure when the situation requires it.

5.2 THIGMOTROPIC RESPONSES

Differential growth in response to a contact stimulus can occur in many rapidly growing organs. For example, if etiolated seedlings are repeatedly touched on one side of the hypocotyl, or if coleoptiles are stimulated by friction along one side, they usually bend towards that side (Bünning 1959a, Pickard 1985a). However, it is not clear whether this behaviour represents a truly tropic response, involving changed patterns of growth in regions other than those directly stimulated, or whether it simply results from a thigmomorphic inhibition of growth only in those regions that receive physical contact.

The stamens of *Portulaca grandiflora* show a thigmotropic response (Jaffe *et al.* 1977). This is a fairly rapid movement, occurring within a few seconds, and results from the loss of turgor and consequent contraction of the cells on the stimulated side of the stamen. [Like heliotropism, (Section 4.1.5) this is a situation in which tropic behaviour results wholly from changes in cell turgor rather than growth.]

The pollen tube in *Lilium longiflorum* shows weak thigmotropism (Hirouchi & Suda 1975). When the pollen grains are germinated *in vitro*, the germ tubes grow randomly on agar but unidirectionally on nylon mesh. The unidirectionality is related to the size of mesh, being greatest on that which seems to give the greatest regularity of mechanical contact. (However, overall, the direction of tube growth seems to be much more affected by the presence of chemical substances from the stigma, than by the mechanical structure of the substratum.)

Directional responses to contact that clearly involve controlled changes in growth patterns are most apparent in the hyphae of some fungi and in the climbing tendrils and the roots of higher plants.

5.2.1   *Fungi*

In certain leaf pathogens, the direction of germ tube emergence and of hyphal growth are regulated by contact stimuli (Carlile 1975). For example, in *Botrytis squamosa* and direct-penetrating fungi, the hyphae grow towards and along epidermal cell wall junctions (Callow 1984). In rusts (*Puccinia* spp.) that penetrate through the host stomata, the uredospore hyphae grow transversely across the surface ridges of the leaf (Callow 1984, Jaffe 1985), such behaviour presumably increasing the likelihood of the hyphae growing across one or more of the longitudinally arranged rows of stomata in cereal leaves.

The mechanical stimuli regulating these responses are thought to be the surface features of the leaf, such as cell walls, cuticular ridges, or surface waxes. Several studies have indicated that inert leaf surface replicas are also effective in regulating the direction of hyphal growth

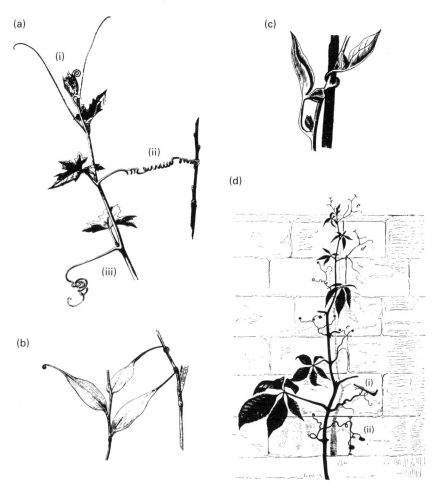

Figure 5.1 Various types of tendrils: (a) *Bryonia dioica*, showing (i) uncoiled tendrils, (ii) free-coiling, (iii) age-coiling; (b) *Gloriosa superba*, with leaf-tip tendrils: (c) *Solanum jasminoides*, with petiole tendrils; (d) *Amelopsis hederacea*, with (i) coiling tendrils, (ii) adhesive pads at the ends of tendrils (from Pfeffer 1906).

(references in Callow 1984, Jaffe 1985). It has been suggested that the repetitive contact stimuli by particular surface features are somehow translated into particular patterns of microfibril orientation in the hyphal cell wall, thus determining the direction of hyphal growth (Dickinson 1977). However, it has also been pointed out that these situations presumably involve the interaction between large polymeric molecules from host and pathogen and that the distinction between mechanical and chemical forms of stimulation may not be absolute (Callow 1984).

[In many rusts, the development of appressoria also seems to be

145

determined primarily by contact stimuli (Wynn 1976), thus constituting a thigmomorphogenic response in fungi. When a hypha reaches a stomatal pore, the surface topography of the pore lips seems to act as the signal that triggers differentiation into an appressorium. However, again the precise surface feature that acts as the stimulus is unknown.]

5.2.2 Tendrils

Tendrils are long slender structures that are adapted to support the plant by attachment to some fixture, usually through some kind of contact-induced coiling response. They vary greatly in size and form, e.g. from a length of around 4 cm in *Bignonia unguis* to 40 cm or more in *Vitis vinifera*, and represent adaptations of virtually all types of organs (Pfeffer 1906). Some of these forms are shown in Figure 5.1, and are discussed further in Box 5.2.

Over the course of their development tendrils can show four different types of growth movements (Satter 1979, Jaffe 1985). During their young unattached stage they often exhibit wide, sweeping ellipses. These so-called 'searching movements' are based upon exaggerated forms of circumnutatory growth, although it has recently been demonstrated that, at least in some species, they may also involve rhythmic changes in cell turgor (Millet *et al.* 1987). In pea tendrils circumnutation always occurs with the long axis of the ellipse at right angles to the sun, possibly through a temperature effect on growth (Jaffe 1980). When the tendril reaches a particular stage of development, contact between its mechanosensitive region and a solid object usually results in the cessation of circumnutation and the initiation of a movement known as 'contact coiling'. The exact form of this contact response varies according to the species (see below), but it results in the tendril becoming wrapped around the stimulating object. The third form of movement shown by the tendrils of some species is called 'free coiling'. This occurs some 15–30 minutes after the contact response and involves the bridging middle region of an attached tendril becoming coiled (Fig. 5.1a), first in one direction, and then in the other when the internal tensions become too great. Free coiling serves the adaptive functions of not only pulling the plant closer to its support, but of also acting as a spring-like buffer against the effects of wind. Further activity within the coiled tendril includes the laying down of some strengthening tissue; tendrils of *Passiflora*, for instance, can individually support over 500 g, which is equivalent to several metres of plant material. Finally, in many species if a tendril never comes into contact with an appropriate stimulus it eventually exhibits 'age coiling', in which loose irregular loops are formed (Fig. 5.1a).

Tendril coiling is thigmonastic or thigmotropic according to the species (Bünning 1959a). Thigmonastic coiling, in which coiling can only occur in one particular direction, is characteristic of tendrils which show a marked dorsi-ventrality in their structure, and generally occurs towards the abaxial surface. (In much of the early literature on tendril coiling the abaxial surface of an organ was described as the 'ventral' side, used in the zoological sense of 'lower' or 'underside'; similarly the adaxial surface was often termed the 'dorsal' side.) Within the thigmonastic class of tendrils, some, such as those of *Momordica*, show a coiling response if touched on any side. Other types of thigmonastic tendrils, however, must be touched on a particular side in order to stimulate coiling, e.g. the abaxial (ventral) side in *Pisum*. (But an appropriately timed touch on the adaxial side of a *Pisum* tendril can negate any previous stimulation of the abaxial side; therefore, in a sense, both sides of a pea tendril are sensitive to contact.) In thigmotropic coiling, as in *Cobaea scandens*, the tendrils show contact coiling towards whichever side is touched.

Mechanosensitivity develops as a young tendril matures, reaching a maximum in *Pisum*, for example, when the tendril is about three-quarters of its final size; at this stage in *Pisum* a small terminal hook also develops on the tendril, curving downwards in the direction of the abaxial, mechanosensitive, potentially concave side of the organ (Jaffe & Galston 1968). Generally, the apical one-third of a tendril is the most sensitive region, and the contact stimulus must be a solid object; water and water drops, e.g. rain, do not provoke a response no matter how forcefully applied. (In the laboratory, a few rubbing strokes with a glass rod is usually enough to induce coiling; on the other hand, unwanted coiling can be avoided by handling the tendril with gelatin-coated rods.) If physical contact with the stimulus is discontinued at an early stage, the tendrils of most species uncoil, presumably through some sort of autotropic straightening response (see Section 2.1). Tendrils are thus unlikely to become attached to supports which themselves show some form of movement or instability.

The rate of coiling response varies greatly according to species. For example, coiling is apparent in *Passiflora* within 30 seconds of stimulation and within a minute or so in *Pisum*, but takes several hours to develop in *Corydalis*. In the species that show rapid responses, coiling results from the contraction of the eventual concave side of the coil and extension of the convex side. In fact in these cases, tendril coiling seems to be a form of movement that involves both turgor responses and growth responses (see Section 5.3.2).

5.2.3 Roots

Roots are highly sensitive to contact and seem to utilize thigmotropic responses to grow around obstacles. Darwin (1880) carried out many experiments on this aspect of root behaviour in a wide variety of species.

In his initial experiments he simply touched the roots, but soon realized that a more continuous form of mechanical stimulation is required to induce a significant directional response. He therefore developed a technique in which small pieces of paper or card (1.25 mm square and 0.15–0.2 mm thick) were more permanently attached to particular regions of the root, and by this means he established that the tip of the root is negatively thigmotropic, e.g. the root of *Vicia faba* bent away from pressures caused by a minimum weight of around 0.32 mg applied to the tip. Darwin concluded that this type of response enables the root to follow the line of least resistance through the soil (see also Box 5.1). However, he also noted that the region behind the tip is positively thigmotropic, and suggested that these two forms of thigmotropic behaviour would together result in the root 'crawling' over soil particles and thus maintaining nutritional contact between the root hairs and the substratum.

The American physiologist Spalding (1894) later described similar behaviour in roots in response to injury, i.e. curvature away from an injury near the tip of the root but towards the side that was injured when the site of injury was further back from the tip. Spalding, in fact, attributed Darwin's results also to injury, in this case perhaps caused by the adhesive that Darwin used to attach his various contact stimuli to the roots. But this interpretation seems unlikely in view of Darwin's finding that when two paper squares of different consistencies were attached to either side of a root tip (each by the same type of adhesive), the root curved away from the paper of greater stiffness.

In many circumstances, thigmotropism seems capable of overriding the strong gravitropic responses even of primary roots. Darwin found that in a vertical bean root, a contact stimulus could divert the root away from the vertical, i.e. thigmotropism overrides gravitropism; but in a horizontal root downward curvature always occurred, even against a contact stimulus, i.e. gravitropism overrides thigmotropism. Pea roots seem to be more thigmotropically responsive, however, and he found that they generally curved away from a contact stimulus, no matter what orientation they were in. Edwards & Pickard (1987) also noted that friction overrides gravitropism in roots.

This interaction, or 'cross-talk' (Pickard 1985a) between thigmotropism and gravitropism may offer important clues to the respective mechanisms of stimulus reception, and these aspects are considered further in Section 5.3.2.

Box 5.2   The climbing habit

In their continual upward growth towards the light, the aerial parts of higher plants eventually require some form of support, and there are two quite different ways of obtaining this:

(a)   Self-support by the development of woody, strengthening tissues; this, however commits considerable amounts of energy and resources to the role of simply keeping the plant upright.

(b)   Support by attachment to some other structure; this so-called 'climbing habit' has the adaptive advantage of enabling the plant to attain maximum height for the minimum investment of resources in supportive tissues.

There are, in fact, a great number of mechanisms by which plants climb, but they can be broadly classed into three general types: scrambling, twining, and the use of tendrils (Darwin 1875a, Pfeffer 1906, Gorer 1968).

In scramblers, the plant becomes loosely and passively attached to another object by some kind of projecting structural adaptation, often a hook or a barb (e.g. climbing roses, blackberry) though some texts also class the adventitious roots of ivy as scrambling devices. This method is thus fairly straightforward, though there are some points of special physiological interest (Pfeffer 1906), e.g. hooks can develop in response to a contact stimulus (*Uncaria ovifolia*), and some species are scramblers in shade conditions but show erect forms in open ground (*Polygonum aviculare*, *Galium mullugo*).

In twiners, the stem of the plant coils around a support. Usually there is no active response or any contact stimulus involved, and twining is simply the result of an exaggerated nutational form of growth. It is therefore more effective on vertical than on horizontal supports, and on supports that are neither too thick nor too slender. (The dodder plant, *Cuscuta* sp., is unusual in that the twining stems are sensitive to contact.) In most twiners, coiling is in a left-handed (anticlockwise) direction, but in hops (*Humulus*) and honeysuckle (*Lonicera*) it is right-handed. In a few species the direction varies from plant to plant (*Polygonum convolvulus*) or even within a single plant (*Tropaeolum tricolorum*, *Ipomoea jucunda*).

Tendrils are characterized by the fact that the coiling reaction is induced by a contact stimulus. However, within this general mechanism, species show great variety in the type of organ that constitutes the tendrils, in the structure of the tendrils, and even, as indicated in the main text, in the details of tendril action. Tendrils can be adaptations of stem (*Vitis*), branch (*Antirrhinum*), leaf (*Gloriosa*, *Lathyrus*), leaflet (*Pisum*), petiole (*Tropaeolum*, *Clematis*), and even root (the aerial roots of *Vanilla* show a weak coiling response to contact). Tendrils can be branched (*Pisum*) or unbranched (*Cucurbita*, *Bryonia*) and often show additional features that aid attachment; the mature tendrils of *Pisum*, for example, have a small hook at their ends, those of *Cobaea* a cluster of claws, and in *Amelopsis* (Virginia creeper) the contact stimulus also induces the formation of adhesive pads at the ends of the tendrils. Some of these features of tendrils are illustrated in Figure 5.1.

5.3   STIMULUS RECEPTION AND TRANSFORMATION

The responses of certain ciliated protozoa to contact may serve as a useful introduction to possible events in other mechanostimulated cells. For example, if *Paramecium* is touched at the front of the cell, the organism backs up then moves off again, usually in a different direction; this is known as the 'avoidance reaction'. If the rear of the cell is touched, the organism speeds up its forward rate of movement, the 'escape reaction'. Both of these responses are initiated by the distortion of the cell membrane, which seems to result in the opening of particular ion channels. The avoidance reaction is characterized by cell depolarization, due to an inward flow of current that is based upon an influx of calcium. The high calcium concentration brings about a reversal of ciliary beat (backward movement); this excess calcium is then pumped out of the cell and the ciliary beat reverts to normal (see Naitoh (1984) for details). The escape reaction on the other hand, involves cell hyperpolarization due to an outward flow of potassium ions, and this brings about a transient increase in the rate of ciliary beat. The difference in the form of response according to which end of the cell is touched, seems to be due to differences in the distributions of specific ion channels in the membrane, i.e. calcium channels towards the front of the cell, potassium channels towards the rear.

In plants, investigations into the mechanisms of reception and transformation of mechanostimuli have naturally enough been largely concerned with the more obvious movements of certain specialized organs. These generally involve changes in cell turgor rather than growth, but they are considered briefly first in order to get some idea of the kinds of events that may be involved in thigmotropic growth responses.

5.3.1   *Events in mechano-induced turgor responses*

General overview
In some mechano-induced turgor movements the stimulus is received by the responding motor cells themselves. In others it is received in special receptor structures or cells, which then send out a signal to bring about the turgor change in the motor cells. This signal is generally in the form of an electrical impulse or action potential (Section 2.1).

Touch sensitivity in plants is considered to reside in the epidermis, and, as in protozoan cells, reception of the stimulus is thought to involve deformation of the cell membrane. In specialized plant organs, various types of structural modifications seem to be involved in helping to bring about this deformation more readily, thus increasing the touch sensitivity of the organ. It is at this level that distinction may occur in the reception of different forms of mechanical stimuli, e.g. certain structural

(a)

(b)

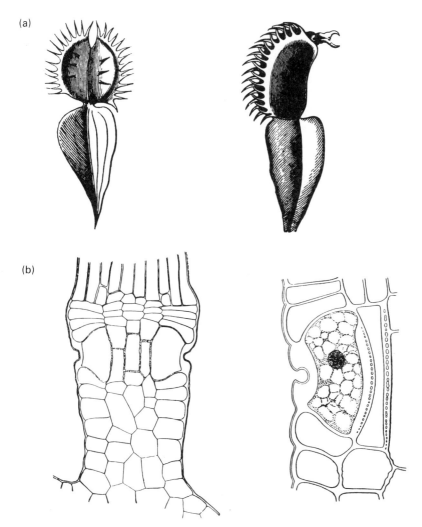

Figure 5.2 The insectivorous plant, Venus flytrap: (a) the trapping lobes in the open position (showing three touch-sensitive hairs on each lobe), and in the closed position; (b) longitudinal sections through the middle and basal sections of a touch-sensitive hair, in the normal position (left) and undergoing mechanostimulation (from Haberlandt 1914).

features may favour the detection of localized pressure differentials that are generated by contact (thigmostimuli), while not being particularly sensitive to vibration or flexure (seismostimuli). But however it is brought about, the deformation of the membrane seems to result in a change in its ionic permeability and thus in the generation of the regulatory action potential.

Venus flytrap

Touch sensitivity in the insectivorous plant *Dionaea muscipula* is most apparent in the three tactile hairs on each lobe of the trap (Fig. 5.2a). The trap is triggered when two hairs are touched or when one hair is touched twice within a few seconds (a simple means of avoiding closure in response to false alarms). When a hair is appropriately stimulated, an action potential of about 50 mV is generated in the base cells of the hair, and then transmitted across the lobe tissues to trigger turgor loss in the motor cells along the hinge of the trap and thus bring about trap closure, all within a fraction of a second (Pickard 1973, Simons 1981, Findlay 1984, Jaffe 1985). The action potential is based on a calcium flux (Hodick & Sievers 1988) and is transmitted at around 200 m s^{-1}.

It has recently been shown that all the cells of the trap lobes are as electrically excitable as the hair cells, and that the high mechanosensitivity of the hair is therefore due simply to its morphology rather than to any biochemical specialization of its membranes (Hodick & Sievers 1988). Each tactile hair shows three structurally distinct regions: the tip, the middle and the base. The middle region consists of wedge-shaped cells whose walls become deformed when the hair is touched (Fig. 5.2b). Haberlandt (1914) first pointed out that the tip region of the hair thus acts as a 'tactile lever' in magnifying the effect of touch on the deformation of the wedge cells in the middle region. It is the deformation of the membrane of these wedge cells that generates the receptor potential. Accumulation of enough receptor potential then brings about the depolarization of the base cells of the hair and the generation of the action potential. The action potential presumably induces some change in the membranes of the motor cells of the trap, which results in their sudden loss of turgor and consequent collapse.

Sundew

The insect-trapping leaves of *Drosera rotundifolia* carry many tentacles which curve inwards in response to touch (Fig. 5.3a). This is a relatively slow movement, taking about 3 minutes to complete, but sticky mucilage on the tentacles hinders the escape of the insect. In fact, the early investigators described this movement as a growth response rather than a turgor response, and Pickard (1980) has pointed out that, if this is the case, then this would represent at least one situation in which an action potential seems to be involved in the regulation of growth. [Some workers (e.g. Lloyd 1942) considered the response to be chemotropic, but that may only apply to the later movements of tentacles that have not received mechanical stimulation, once an insect has been trapped.]

Mechanosensitivity resides in the cells at the head of each tentacle (Fig. 5.3b). The head cells are characterized by structural features known as 'tactile pits' (Haberlandt 1914). These are exceptionally thin areas of the

(a)

(b)

(c)

10μ

Figure 5.3 The insectivorous plant, sundew: (a) a leaf trap carrying many touch-sensitive tentacles; (b) a single tentacle showing the touch-sensitive head cells; (c) a single protoplast isolated from one of the head cells, showing protoplasmic projections that penetrate the pits on the outer cell wall (from Haberlandt 1914).

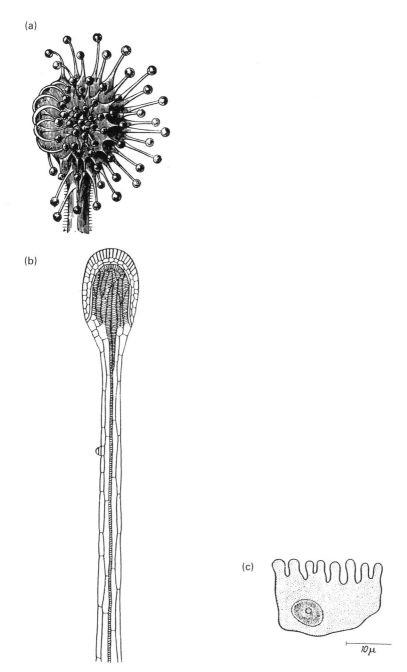

cell wall which are filled with closely adpressed projections of the protoplasm (Fig. 5.3c). The cytoplasm-filled cavities in the wall may represent another type of structural adaptation which enables the cell membrane to be more easily deformed by contact or pressure (Haberlandt 1914). Alternatively, they may act to provide closer electrical continuity between the external mucilage and the interior of the head cell.

When an insect touches the head of the tentacle, the electrical potential of the mucilage is lowered (Pickard 1973). This effect is thought to act as the receptor potential, and to generate the action potential of about 20 mV, which is transmitted to the base of the tentacle at around 5 mm s^{-1}. Exactly how this brings about curvature, and which are the responsive cells, is not clear.

Reproductive organs

The touch-sensitive stamens, styles, or other organs that operate in the flowers of many species (see Table 1.2) also show the kinds of structural adaptations and electrical responses that seem to characterize mechanostimulation. In the stamens of *Portulaca grandiflora* the epidermal cells carry distinct dome-shaped structures called 'tactile papillae' (Fig. 5.4a, b, c), which are capable of being visibly deformed by contact or pressure (Haberlandt 1914, Jaffe *et al.* 1977). And the generation of action potentials in response to mechanostimulation has been observed in several types of reproductive organs, such as the stigmas of certain of the Scrophulaceae and the stamens of *Berberis* and *Opuntia* (Pickard 1973).

The sensitive *Mimosa* plant

The general features of the turgor movements shown by *Mimosa pudica* have been described in Section 1.2.2. The turgor changes in the motor cells of the various pulvini can be brought about by an extremely wide variety of stimuli, including mechanostimuli of contact, pressure, flexure, and vibration. (Other types of stimuli that are effective are: a sudden temperature change of as little as 5 °C; high intensities of light, with greatest effectiveness in the blue and the red regions of the spectrum; toxic chemicals, such as ether, chloroform, alcohols, ammonia, and acids; and any form of injury.)

The exact means by which the mechanical stimuli are detected is not entirely clear. The only obvious structural feature that seems relevant to this aspect is the high density of surface hairs. Haberlandt (1914) suggested that these may magnify the effects of mechanical stimuli on the excitable cells of the pulvini themselves.

However, the outstanding and, in the context of this section, illustrative feature of *Mimosa pudica* is the extent to which action potentials are generated and transmitted throughout the plant in response to stimu-

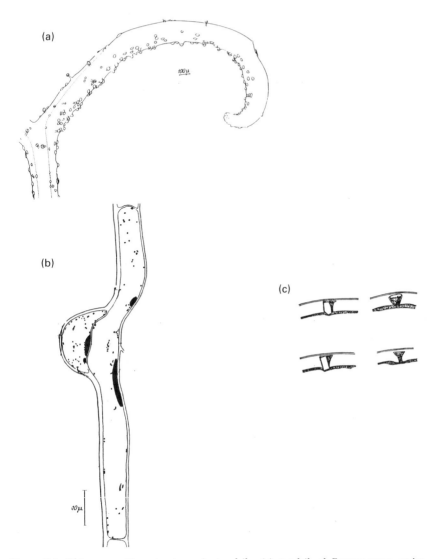

Figure 5.4 Thigmosensitive structures in tendrils: (a) tendril of *Eccremocarpus scaber*, showing the general surface distribution of tactile papillae; (b) longitudinal section through an epidermal cell and its associated papillus (from Baillaud 1959); (c) longitudinal section through an epidermal cell of a tendril of *Cucumis*, showing a tactile pit and its pressure wedge of a crystal of calcium oxalate (from Haberlandt 1914).

lation by mechanical and other means. These various action potentials all activate the pulvinar motor cells to bring about stem and leaf folding (see Fig. 1.2), but certain other features differentiate them into several distinct types (Roblin 1979):

(a) 'Middle wave' types of action potential are generated by non-injurious stimuli; these are propagated through the tissues at rates of 50–500 mm s^{-1}, possibly travelling in the phloem; they do not pass through dead tissues and are blocked or dissipated at the first pulvinus that they come to.

(b) 'Slow wave' action potentials are generated in response to injury; these travel at around 5 mm s^{-1}, possibly in the xylem, and they can pass through pulvini, dead regions, and water gaps. This type of action potential seems to be associated with some chemical known as 'wound hormone' (or Ricca's factor), which may be related to the recently discovered turgorins (see Section 1.2.2). The significance of this type of behaviour to our general consideration of mechano-stimulation is that it indicates that a close interrelationship can exist between electrical action potentials and regulatory chemicals.

(c) 'Rapid wave' action potentials are generated by very serious injury; they are transmitted at rates of 1000–4000 mm s^{-1}, but their basis is completely unknown.

5.3.2 Events in mechano-induced growth responses

Tendrils

The epidermal cells in mature tendrils generally show some form of structural modification that seems likely to facilitate the effects of contact or pressure on the cell membrane (Fig. 5.4). These structural features can be of various types depending on the species (Haberlandt 1914). In many of the Cucurbitaceae tactile pits are present in the walls of the epidermal cells, often with a crystal of calcium oxalate embedded in the pit (Fig. 5.4c); this presumably acts as a 'pressure wedge' to increase still further the mechanosensitivity of the epidermis. In *Luffa*, however, warty protrusions are present on the epidermis. And in *Eccremocarpus scaber* the epidermal cells of the tendrils carry tactile papillae. Tendril epidermal cells also seem to have an unusually high number of plas-modesmata (Junker 1977a), suggesting that the whole surface tissue acts as a physically unified system (Satter 1979). The distributions of all these types of structural features generally correlate well with regions of high mechanosensitivity on the tendrils, being more numerous, for example, on the abaxial sides and towards the tips in tendrils of *Eccremocarpus* and *Luffa* (Junker & Reinhold 1975, Junker 1977a). However Pfeffer (1906) noted that structural modifications were also present on the *in*sensitive regions of *Bryonia* tendrils, indicating that additional physiological features may be involved in mechanosensitivity.

Action potentials have been detected in cucurbit tendrils within seconds of thigmostimulation (Pickard 1973). These potentials are of the order of 60–100 mV, and are propagated along the tendril at rates of around 0.4 mm s^{-1} (Jaffe 1985). However, such action potentials may only occur in those types of tendrils that show the rapid 'contact coiling' form of response, e.g. action potentials were not detected in the much more slowly coiling tendrils of *Parthenocissus* (Jaffe 1985).

At the cellular level, two types of response seem to occur during coiling, although the relationship or interaction between them is not entirely clear. In tendrils of *Pisum*, the rapid contact coiling response is due to changes in cell turgor; the cells on the potentially concave (abaxial) side of the tendril contract, while those on the convex side expand (Jaffe & Galston 1968, Jaffe 1985). The coiling induced by these initial events is then reinforced by differential growth responses, with the convex side of the tendril elongating faster than the concave side. [Different forms of coiling, e.g. slow coiling in other species, 'age coiling', or hormonally-induced coiling (see later), may only involve the differential growth responses.]

Another significant regulatory factor in coiling is light, at least in the responses of *Pisum* tendrils. If these tendrils are thigmostimulated in darkness, coiling does not occur until they are exposed to light, and in fact the mechanosignal seems able to be 'stored' within the tendril in darkness for 90 minutes or longer (Jaffe 1985). This effect has been suggested to be due to a requirement for ATP for contractile coiling, but it may equally represent a photomorphogenic effect of light on the growth responses of the tendril. [The former suggestion is related to the speculation that a contractile ATPase may be involved in ion movements during the turgor responses of coiling (Jaffe & Galston 1968), but this aspect has not been developed further.]

The regulatory mechanisms responsible for coiling are not yet clear. The initial turgor changes of contact coiling may be induced by action potentials and the later growth responses controlled by regulatory chemicals. Alternatively, the relationship between these mechanisms of regulation may be closer, and the action potential may act upon some aspect of hormonal behaviour.

The nature of any hormonal involvement in coiling is also not understood. Exogenous application of IAA usually increases the rate of coiling, and in some cases can even bring about coiling in the absence of any contact stimulus (Reinhold *et al.* 1970, Jaffe 1975, Junker 1977b). (High concentrations of carbon dioxide can also initiate a coiling response, and ethylene promotes coiling in some species.) However, the differential growth of coiling is not simply due to the establishment of a lateral gradient of auxin across the tendril. In the first place, auxin stimulates coiling when it is supplied symmetrically as a bathing solution (Reinhold *et al.* 1970). And secondly, experiments with radiolabelled IAA indicated that a symmetrical distribution of radioactivity is maintained in a touch-stimulated tendril both before and during coiling (Junker 1976, 1977b). Further circumstantial evidence against auxin being the primary messenger in the coiling response is obtained from a comparison of the characteristics of auxin movement and coiling development: auxin transport takes place in a basipetal direction, but coiling develops both acropetally and basipetally from the site of stimulation (Junker 1977a); and in tendrils auxin is transported at 14 mm h^{-1} (Junker 1976), but the physiological stimulus for coiling seems to move along the tendril at 3 mm min^{-1} (Junker 1977b). Therefore the effect of auxin on coiling is considered to be due to some previously established difference in sensitivity to auxin between the two sides of the tendril (Jaffe 1985).

The basis of any such differential sensitivity to auxin in a stimulated tendril is unknown. It may derive from differential exposure of auxin-receptor sites on the membranes of the growing cells (Jaffe 1975). That is, the effects of a contact stimulus may bring about changes in the properties of the cell membranes, which not only influence the ionic fluxes of turgor responses but also influence the ability of a growing cell to respond to plant growth regulators.

Roots

Perhaps unsurprisingly, the region of thigmosensitivity in roots seems to be the root cap. For example, the rapid slowing of growth that occurs within 10 minutes or so of the root tip coming into contact with a solid object, does not take place in the absence of the root cap (Goss & Russell 1980). [The cap is therefore far from being simply a protective organ, and there is some justification for the view that 'there is no structure in plants more wonderful, as far as its functions are concerned, than the tip of the radicle' (Darwin 1880); the root cap is the site of thigmotropic and gravitropic sensitivity, and is also a source of growth-regulatory chemicals (see Ch. 3); in general, the cap is renewed every 6–9 days (Feldman 1984), although in a few species it may have a higher rate of turnover.]

Consideration of the wide range of responses that are induced in roots by mechanical stimulation, and of their rates of onset, has led to the view that they must have a hormonal basis, rather than being due simply to the physical effects of pressure (Russell 1977). In roots subjected to impedance by being grown in compacted soils, the levels of abscisic acid do not seem to change but the concentration of auxin in the root tips shows a threefold increase (Lachno et al. 1982). These authors considered this increase in auxin to be responsible for the decreased extension growth and increased development of lateral roots in impeded roots. Increased production of ethylene has also been implicated as a significant factor in responses to mechanical stress in many organs, including roots (Feldman 1984, Pickard 1985a). However, it is still not at all clear how a directional contact stimulus on the root cap is transformed into a directional thigmotropic growth response in the elongation zone.

An intriguing new line of enquiry into the general detection and response to mechanostimulation at the cellular level has recently been developed (Pickard 1985a, Edwards & Pickard 1987). By analogy with the situation in many animal cells, Edwards & Pickard suggest that there may be some form of 'stretch receptors' in the plant cell membrane. These are envisaged to be stretch-activated ion channels, perhaps inter-linked by relatively rigid spectrin-like filaments whose presence would thus act to focus or amplify the effects of a mechanical stimulus in deforming the membrane. Changes in the permeability properties of these mechanotransductive ion gates, possibly relating to the regulatory calcium ion, may then lead to changes in growth or in some aspect of growth-regulator activity.

In support of their model, Edwards & Pickard (1987) review the general occurrence of action potentials and voltage transients, indicative of ion fluxes, that are induced in plant cells by a wide variety of stimuli that seem to act by deforming the cell membrane (including the experimental deformation of the membrane in protoplasts by means of suction). These authors also point out that although responsiveness to mechanical stimulation may thus be a general property of all plant cells, the structural adaptations of some cells, e.g. those in touch-sensitive organs, increase the mechanosensitivities of their membranes. In fact Edwards & Pickard (1987) note that:

. . . the general concept (of mechano-sensitivity in plant cells) was put forward long ago by Wilhelm Pfeffer, whose study of plant movements convinced him that 'Mechanical agencies probably awaken more or less feeble reactions in all plants . . . ' and that 'The shape and relationships of the cell and cell-wall, as well as the way in which the cells are joined and arranged in the tissues, may render the perception of the stimulus more readily possible at particular points, but do not produce this special form of irritability.' (Pfeffer 1906, p. 66).

General remarks
The above quotation from Pfeffer also indicates the possible wider implications of mechanosensitivity in plant cells. External sources of various types of mechanostimuli affect several aspects of plant growth and development, and, as we have seen, different types of structural adaptation seem to be involved in increasing the sensitivity towards particular types of stimuli; in these cases, the adaptations are located externally to the membrane. Furthermore, mechano-induced changes in the properties of the cell membrane may also be involved in the detection

of gravitropic stimuli (Pickard 1985a, Edwards & Pickard 1987), either by direct, gravitationally induced deformation of the membrane (Dennison 1971, Clifford *et al.* 1982) or through the intervention of statoliths; in the latter case, the structural adaptation responsible for increasing the sensitivity of the detection system is located internally to the membrane.

However, in addition to these responses to externally applied mechanostimuli, mechanically induced changes in the membrane may also occur as a result of the internally generated forms of mechanical stimulation that must arise from the forces and pressures developed within growing tissues (see also Box 3.2). It has long been recognized that mechanical stress, both internally generated or externally imposed, is a potent morphogenic factor, and a growing organ has been described as 'a mechanical equilibrium of mutually opposing tensions and compressions' (Lintilhac & Vesecky 1981). In the past, much attention has been given to the longer-term effects of such forces on cell division and wall orientation (see Lintilhac & Vesecky 1981, 1984). However, these 'tissue-generated' forces are also likely to be operating on rates of cell elongation in the shorter term. It may be that mechano-induced changes in membrane properties, brought about by internally generated forces, underlie such endogenously regulated growth responses as circumnutation, epinasty, and autotropic straightening reactions.

CHAPTER SIX

Other Tropisms

Several other environmental factors besides light, gravity, and physical contact can influence the orientation of plant organs, and responses to these factors have received such self-explanatory names as chemotropism, hydrotropism, traumatropism, electrotropism, and so forth. Some of these responses are highly significant to particular organisms at particular stages of development, and have therefore been intensively investigated. Others, however, exist largely as descriptions in the older literature and in many of these cases there may be some doubt about the 'tropic authenticity' of the response. In the first place, early observations were often carried out under relatively uncontrolled conditions, or under conditions in which the potential influence of other regulatory factors was unappreciated. (For example, some of the early descriptions of hydrotropism were made on material responding under unspecified light conditions.) Secondly, some of the early reports might arguably be debarred from inclusion as tropic responses because of the lack of specificity in mechanisms or in the way in which the direction of growth is influenced. (For example, should the localized effect of some toxic factor be classed as a tropic response even though it brings about a change in the direction of growth?)

In some of the situations that are considered in the following sections, it may be useful to keep in mind the characteristic features of a tropism (see also Section 2.1):

(a) It is initiated by a specific stimulus, which itself is strongly directional.

(b) The stimulus is (usually) detected by a specific receptor, whose action evokes a specific transduction sequence.

(c) The response involves a change in organ orientation, the direction of which is specifically related to the direction of the stimulus.

(d) A tropic growth response usually involves stimulation and inhibition, and often occurs in regions of the organ other than those that directly receive the stimulus.

(e) There is usually a non-linear relationship between the level of the stimulus and the extent of the response, and often some form of sensory adaptation to the stimulus.

6.1 CHEMOTROPISM

6.1.1 General introduction

Chemotropism is a directional growth response that is provoked by a chemical stimulus; the term chemotaxis is used to describe the locomotory responses of motile organisms or gametes. Both forms of response can be positive, towards a beneficial or attractive substance, or negative, away from a harmful or unattractive substance.

Responses to chemical substances in the external environment are more common, and more thoroughly investigated, in lower plants, particularly in unicellular organisms and gametes. In multicellular organisms there is generally a greater homeostasis or 'buffering' of the cellular environment, and chemoresponses to external chemicals seem restricted to fairly specialized situations.

A very wide range of chemicals is involved in these types of responses, reflecting the variety of adaptive advantage that can be conferred in these situations. For example, simple nutrients and oxygen can induce positive (attractive) responses in feeding behaviour or host location, and biological waste products such as organic acids, alcohols, phenols, and carbon dioxide can bring about negative (repellent) responses in the avoidance of poor or overcrowded environments. Again, specific chemicals can induce responses only in particular organisms or at particular stages of development; these more specific forms of response commonly occur in the reproductive behaviour of lower plants.

General chemical effects and specific chemoresponses
When considering any putative chemotropic response, three interrelated questions should be asked:

(a) *Is the response brought about by a directional stimulus?* That is, does the response occur in relation to an actual gradient of the chemical, or does it occur simply because of the presence or absence of the chemical? For example, the general stimulation of growth brought about by the presence of a chemical (or absence of an inhibitor) can produce a straightening effect on an organ that, in the short term, can seem to be a directional response; this is particularly the case with simple unicellular structures such as fungal hyphae, pollen tubes, and so forth.

(b) *Is it a truly directional response?* That is, does the response derive from

162

an actual change in the direction of growth towards (or away from) the source of the stimulus, or does it simply represent some form of 'nutritional entrapment'? For example, profuse growth and development (particularly branching) can occur in an organ, or part of an organ, that is in an especially nutrient-rich part of the environment; this biased morphology can hide the fact that the organ originally entered that part of the environment by accident.

(c) *Is the response brought about by a specific chemical?* That is, does the response occur as a result of the specific interaction of a particular chemical with a particular cellular receptor, or does it occur simply because of some disruption of general metabolism? For example, effects on general metabolism, say by a nutrient or by some toxic compound, can produce localized growth stimulation or inhibition that influences the overall direction of growth of the organ.

Certain investigative approaches can be useful in determining whether the response is directional and specific. The questions of directionality can be answered by the simple but important test of repositioning the source of the chemical or test factor and noting whether this brings about an appropriate change in the direction of growth of the organ (for example, see Fig. 6.3). Questions about specificity are not quite so straightforward to answer. Some preliminary indication about the likelihood of the response being due to general or specific effects may be given by the nature of the chemicals themselves, e.g. factors that might be expected to act through general metabolic effects include nutrients (though many of these do act specifically), ions (especially protons, potassium, calcium, and magnesium), acids, alkalis, heavy metals and metabolic poisons, oxygen and carbon dioxide levels, water availability, and temperature. Secondly, chemicals which affect the direction of growth in some specific ('non-metabolic') fashion might be expected to do so without themselves being metabolically altered, in which case sterically related, but non-metabolized, analogues of the chemicals might also be effective in influencing the response. (This is the case in specific chemotactic responses of bacteria, see below.) And thirdly, if a chemical is having a specific effect on growth it must be interacting with a specific receptor, which is likely to be located on the cell membrane; therefore, chemotropically effective chemicals, and their analogues and competitors, are likely to bind to the membrane or a membrane component, and the specificity and kinetics of such binding should match the specificity and kinetics of chemotropic action. By this stage of the investigation of course, little doubt would remain about the specificity of action of the chemical, and study of the actual mechanism would be well underway.

(a)   (b)

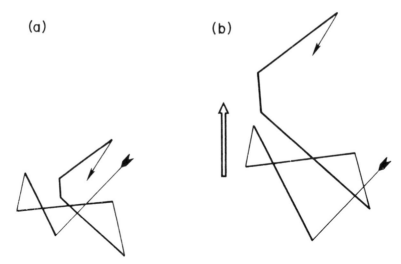

Figure 6.1 Diagrammatic representation of the swimming trajectories of a bacterium. (a) Normal 'random walk' in a homogeneous medium, resulting from alternate swimming and tumbling. (b) In a concentration gradient of a chemoattractant (open arrow), the durations of swimming periods in an up-gradient direction (heavy lines) are extended because of tumble suppression, resulting in migration up the gradient in a time-biased random walk. (From MacNab 1979.)

Methods of detection of a chemical gradient

There are two different methods of detecting the direction or gradient of a stimulus, known as temporal sensing and spatial sensing (Section 2.1). Temporal sensing is used by small, locomotory organisms, usually unicells, and involves the comparison of stimulus level (concentration of chemical) at one point in time with the level at another point in time, and the induction of some behavioural response if there is any change in level. Mechanisms involved in this form of chemosensing have been particularly well studied in the chemotactic responses of bacteria, and a brief outline of these may provide a useful basis for the subsequent consideration of responses in the cells and gametes of lower plants.

Peritrichous bacteria alternately swim (by counter-clockwise rotation of the flagellar bundle) and 'tumble' (by clockwise flagellar rotation). The swimming periods result in locomotion of the cell in a straight line and the tumbles in a random directional change (see MacNab 1979, Taylor & Panasenko 1984). Locomotion therefore follows the pattern of a 'random walk' in three dimensions (Fig. 6.1). A concentration gradient of chemotactic agent affects this pattern by affecting the relative frequency of tumbling. For example, a sudden increase in the concentration of a chemical attractant transiently suppresses tumbling (by interacting with components of the cell membrane and thus influencing flagellar activity, see below); the effects of this tendency to suppress tumbling are therefore to generate a time-biased random walk up the gradient in the direction of increasing concentration (Fig. 6.1b). Conversely, a chemical repellent is effective by enhancing the frequency of tumbling and thus biasing the movement in the direction of decreasing concentration. Note that this mechanism is not responding simply to the presence or absence of the chemical; it is a change in the concentration that triggers a transient change in the pattern of flagellar activity. Therefore it is a chemical gradient that is effective in eliciting the response. (And some form of sensory adaptation to chemical concentration is an integral part of the mechanism.)

Spatial-sensing methods of detecting chemical gradients seem to be used by larger, slow-moving organisms and involve comparisons of concentration in one region of the organism with concentration in another region, although little is known of the details of any spatial chemosensing mechanism. However, these methods are thought to underlie the chemosensory behaviour of leucocytes, and cAMP-induced aggregation of the cellular slime mould *Dictyostelium* (Taylor & Panasenko 1984, Bilderback 1985). In *Dictyostelium*, for example, an amoeba is sensitive to, and orients itself in, a cAMP gradient equivalent to a concentration difference of 3.6×10^{-11} M between one end of the cell and the other (see Bilderback 1985).

The chemosensing mechanisms that operate in plants are almost entirely unknown. By analogy with bacteria and ciliated protozoa, temporal sensing presumably operates in motile unicellular plants and reproductive gametes. But in other organs such as algal filaments, fungal hyphae, and higher plant cells, the nature of the sensing mechanisms is not clear.

Mechanisms of reception and transformation of a chemical stimulus

In non-specialized cells, mechanisms of reception and transduction of a chemical stimulus have received most attention in bacteria and protozoa, and again a brief account of these cases may be useful to give some idea of what may be happening in other cells (see MacNab 1979, Taylor & Panasenko 1984, for details). As in other sensory phenomena, the cell membrane plays a central role.

In bacteria a very wide variety of chemicals can induce chemotactic responses. Many of these operate through non-specific effects on the cell membrane or on metabolism, e.g. pH, Ca^{2+}, Mg^{2+}, O_2. Others however, exert their chemotactic effects directly, without being metabolized or chemically changed. This has been determined not only by investigation of metabolic events, but also by the use of mutant cells which are unable to metabolize the chemicals, and by the use of non-metabolizable steric analogues of the chemicals. These specific chemotactic chemicals can be grouped into several categories (MacNab 1979). There are two general classes of attractants, amino acids and sugars (D-pentoses and D-hexoses). Repellents include fatty acids, aromatic and indole compounds, and other bacterial excretory products. Within these general classes there are several types of 'specificity groups', i.e. groups within which the chemicals cross-interfere and compete in chemotactic effectiveness. The receptor molecules for these specificity groups can be identified and isolated in cell preparations by means of their specific interactions with the chemicals. In *Escherichia coli* there are 25 attractant receptors; the sugar receptors are small proteins of molecular mass 30–40 kDa, which are loosely attached to the cell membrane, and which seem to carry one type of sugar-binding site per molecule. Thus, the general chemotactic mechanism is initiated by the binding of the attractant to its appropriate receptor on the membrane; this brings about a configurational change in the receptor molecule, a consequent change in membrane properties, and a transient change in flagellar activity.

Studies of protozoan cells such as *Paramecium*, indicate that the chemo-induced changes in membrane properties involve specific changes in the activities of calcium and potassium channels. Consequent transient changes in the cellular concentrations of calcium and potassium bring about transient changes in ciliary beat: increased calcium brings about a reversal of beat and backward swimming, decreased potassium speeds up forward motion (see Taylor & Panasenko 1984 for details). Therefore, in these cells at least, the transduction and mediation of chemical stimuli seem to involve the same sorts of events as in responses to contact and mechanostimuli (see Section 5.3). (In fact, in *Paramecium* and *Tetrahymena* the close relationship between the systems involved in chemosensing and mechanosensing is demonstrated by the cross-interference of chemical stimuli and mechanical stimuli in their respective responses, see Naitoh 1984).

6.1.2 Responses in fungi and lower plants

Chemotropic and chemotactic responses are widespread among lower plants and can be categorized into two general types:

(a) 'Nutritional' responses, i.e. responses involved in finding nutrients, energy sources, or a suitable host.

(b) 'Social' responses, i.e. responses involved in non-nutritional interactions between organisms, during vegetative development and especially during reproductive behaviour. Many algae and fungi release chemicals which not only act as attractants between opposite mating types, but also induce and regulate actual sexual development. These latter aspects are touched upon in the following sections, but access to the voluminous literature on these topics is best achieved through consultation of the accounts by Gooday (1975), van den Ende (1976, 1984), Kochert (1978), Callow (1984), Wiese (1984), and Bilderback (1985).

Fungi

Nutritional responses The dependence of fungi upon organic sources of energy may be the reason for the fairly common assumption that their hyphae grow towards nutrients. However, in the case of most fungi, there is little actual evidence of chemotropic behaviour (Gooday 1975). Hyphae do, of course, grow much better in a nutrient-rich environment, and branching is often more profuse, but these simply represent cases of nutrient entrapment, not attraction towards the nutrients. Early work that seemed to indicate chemotropic effects is now recognized to have been due to negative autotropism (see below) or to positive responses to oxygen. Therefore except (perhaps) for positive aerotropism, there seem to be no generally occurring nutritional chemotropic responses in zygomycetes, ascomycetes, or basidiomycetes (Gooday 1975).

Nutritional chemotropism does occur in certain fungi, however. It is a well-established phenomenon in aquatic phycomycetes (which are often considered to be phylogenetically distinct from other fungi). For example, vegetative hyphae of *Saprolegnia* grow towards mixtures of amino acids at concentrations of 2.5 mg ml^{-1}, and those of *Achlya* towards casein hydrolysate at 20 mg ml^{-1} (Gooday 1975, Bilderback 1985). In these cases the phenomena have passed the chemotropic test of showing an appropriate directional change in response to the repositioning of the stimulus (although the high concentrations of attractants that are required suggest that the chemicals act through general metabolic effects rather than through specific receptors).

Many parasitic fungi also show chemotropic responses. In several pathogenic phycomycetes, such as *Phytophthora* and *Pythium*, the biflagellate zoospores are chemotactically attracted towards roots and the eventual germ tubes also show positive chemotropic responses (Callow 1984, Bilderback 1985). The precise nature of the attractant from the root is not clear; a wide variety of amino acids, carbohydrates, and mixtures seem to be effective. The main region of attraction is to the elongation zone of the root (Mitchell 1976). This may be the region that is most active in establishing a gradient of exudate, but it is also the region around which extracellular loops of electric current are centred (see Section 3.2.3), and it has been suggested that electrotactic and electrotropic responses may be involved in the behaviour of root pathogens (Miller *et al.* 1986, 1988).

Again, in leaf-infecting fungi such as the downy mildews, the zoospores and the germ tubes often show strong directional responses towards the stomata (Ziegler 1962a, Callow 1984, Edwards & Bowling

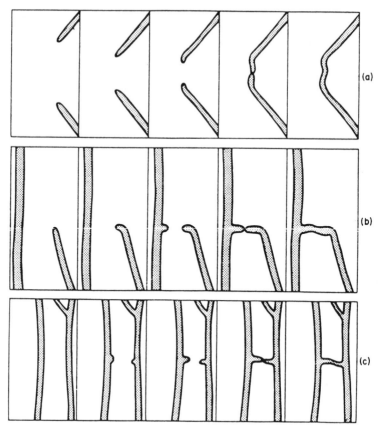

Figure 6.2 Records of successive stages in various forms of fusion in vegatative fungal hyphae: (a) hyphae to hyphae; (b) hyphae to hyphal 'peg'; (c) hyphal 'peg' to hyphal 'peg'; (from van den Ende 1976, after Buller).

1986). These responses have been variously shown to be associated with gradients of organic compounds, carbon dioxide concentration, and pH. (Mechanostimulation from leaf surface structures may also be involved, see Section 5.2.)

Social responses Chemotropic responses in fungi are more common in relation to social behaviour, and include responses both of vegetative hyphae and of reproductive structures. Directional responses of vegetative hyphae of the same individual or of the same species are referred to as 'autotropic' responses. (Note that this use of the term is slightly different from the meaning of autotropism in relation to organs of higher plants, see Section 2.1.1). Fungal autotropism can be negative (i.e. avoidance) or positive (i.e. attraction). Examples of negative autotropism

168

are seen in the circular forms taken up by fungal colonies growing in culture on agar plates, where the hyphae are generally spaced equidistantly from each other whatever the nutritional status of the medium. [It has been suggested (Gooday 1975) that in the past, negative autotropism was mistakenly identified as positive chemotropism; the chemotropic test of repositioning the source of the stimulus distinguishes between the two types of responses.] The basis of this self-avoidance is still unclear. It may be due to the excretion of labile 'staling factors' which influence the direction of growth of adjacent hyphae. Alternatively, some cases of negative autotropism may be due to competition for oxygen, i.e. positive aerotropism (see Gooday 1975).

Examples of positive autotropism between vegetative hyphae are seen in the vegetative fusions that result in an anastomosing, three-dimensional mycelial network, particularly characteristic of ascomycetes and basidiomycetes. In the 1930s A. H. R. Buller catalogued the various forms of such fusions (Gooday 1975, van den Ende 1976), and his carefully recorded observations leave no doubt that some form of attraction is involved (Fig. 6.2). However, the nature of the attractant is still unknown, and even the general form of the mechanism is still unclear; for example, the response cannot be due to just a single chemical, otherwise there would be no gradient between individual hyphae.

Sexual behaviour in fungi is very much controlled by chemicals, and within this area chemical attraction is a major means of bringing together the mating partners (Fig. 6.3). The chemicals involved in these responses are referred to as sex hormones, although the term pheromone is also used (the older term 'gamone' has largely disappeared). As previously mentioned, some sex hormones regulate sexual development as well as serving as chemotropic or chemotactic attractants (see Gooday 1975, van den Ende 1976, 1984, Kochert 1978, Bilderback 1985). For example, in the water mould *Allomyces*, a sex hormone from the female gametes attracts the male gametes. This compound was in fact the first plant sex hormone to be discovered, in the 1930s. It is a bicyclic sesquiterpenediol, although its common name of 'sirenin' seems more biologically descriptive. A response can be induced by a concentration of 10^{-10} M. Sirenin binds to a surface receptor on the male gametes and exerts an effect on flagellar activity so that the gametes undergo fewer directional changes (temporal sensing) and consequently swim up the gradient (van den Ende 1984).

In the unisexual species of the freshwater fungus *Achlya*, sexual attraction and development are co-ordinated by the actions of several steroid-type sex hormones. The female strain produces 'antheridiol' which not only induces the formation of antheridia on the male strain but also chemotropically attracts the growth of the antheridial hyphae towards the female strain. The antheridia themselves produce 'oogoniol'

169

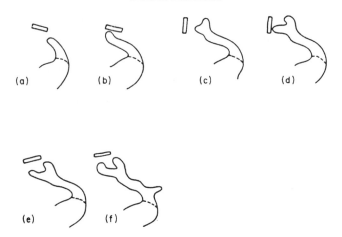

Figure 6.3 Chemotropism during sexual reproduction in the fungus *Ascobolus stercorarius*. (a) Oidium and ascogonium placed in position; (b) 12 min, directional growth of the trichogyne towards the oidium; (c) 25 min, redirectional growth by 'bulging' in response to the repositioning of the oidium at 16 min; (d) 37 min, growth of the lateral branch directly to the oidium, together with a further change in the direction of the original tip; (e) 41 min, oidium again repositioned; (f) 48 min, both apices respond to the new position of the stimulus. (From van den Ende 1976, after Bistis.)

which induces the formation of oogonia on the female. (It is now known that several other chemicals also interact in this system and modify the actions of antheridiol and oogoniol, see Bilderback 1985.)

Sex hormones also act in the responses of many other fungi. In the filamentous species of the Mucorales, carotenoid derivatives known as 'trisporic acids' are involved in the induction of reproductive structures and in the subsequent chemotropic attraction between these structures on different mating strains. In the ascomycete *Ascobolus*, the tip of the young ascogonium is chemotropically attracted towards the antheridium. And in the basidiomycete *Tremella*, the conjugation tubes of different mating types grow chemotropically towards each other. (See Kotchert 1978, van den Ende 1984, Bilderback 1985.)

Little is known about the actual mechanisms involved in the chemotropic growth responses of fungi. Hyphae grow by the activity of a small region at the hyphal tip (Gooday 1975). 'Bulging' type responses, due to changes in the positions of the growing points (see Section 2.1), can be discerned in many of the directional growth changes (e.g. Fig. 6.3). The wide range of chemicals that are usually effective in a 'nutritional' chemotropic response, and the concentrations required, make it likely that these substances act through general effects on metabolism. On the other hand, the sex hormones must act through specific mechanisms, and seem likely to control the position of the hyphal growing point by binding to receptors on the cell membrane. [Action by chemical agents may also be responsible for some cases of fungal 'rheotropism', i.e. directional responses to the flow of water. For example, positive rheotropism (upstream growth) in germ tubes of *Rhizopus nigrans* seems to be due to high concentrations of a growth inhibitor being leached from the downstream sides of the spores (Carlile 1975). Other cases of negative rheotropism may be due to the downstream release of a growth stimulator. Alternatively, such growth effects may arise from mechanostimulation of the cell membrane by flexure, see Section 5.2.]

Algae and lower plants

In green plants, of course, the energy source is usually light, rather than some organic chemical. Gravity and light are therefore the major directional signals involved in the regulation of vegetative growth. However, chemicals still act as an important means of communication between individuals in the reproductive behaviour of lower plants, again as agents which both induce sexual development and serve to attract opposite mating partners.

Descriptions of actual chemotropic responses are relatively rare, although they do seem to be involved in the directional growth of the copulation tubes during filament conjugation in *Spirogyra* (Ziegler 1962a, van den Ende 1976, Wiese 1984). However, the nature of the substances involved, and their modes of action, are unknown.

Chemotactic behaviour in motile gametes is very common (Wiese 1984, Bilderback 1985). In *Chlamydomonas*, unidentified attractants from the gametangia direct the locomotion of the sexual gametes of the opposite mating types. In *Oedogonium*, the oogonia release a substance (MW 500–1500, volatile, heat labile) which attracts the male gametes.

In the brown seaweeds, more is known about the chemistry of these so-called 'erotactins' (Wiese 1984). In *Fucus* and *Ectocarpus*, the chemicals that are released by the non-motile eggs to attract the motile sperm have been identified as specific hydrocarbons (termed, respectively, fucoserraten and ectocarpen) and synthesized in the laboratory. In the presence of the attractants, the sperm stop swimming randomly and travel in circular paths until the source of the attractant is located. Use of radiolabelled ectocarpen has indicated that the attractant binds to the anterior flagella of the sperm, but not the posterior (Wiese 1984, Bilderback 1985).

In ferns, gibberellin-like substances known as 'antheridiogens' are produced by the gametophyte thallus, and induce the formation of antheridia. The motile sperm respond chemotactically to the relatively simple chemical, sodium malate, which seems to alter the locomotory action of the flagella (Bilderback 1985).

In all of these situations, the phenomena of chemical attraction are quite clear, but relatively little is known about the actual mechanisms of stimulus reception and directional guidance.

6.1.3 Responses in higher plants

In higher plants, not only is vegetative growth directionally influenced mainly by gravity and light, but reproductive behaviour is also largely controlled by light and temperature. Chemical signals in the external environment therefore play a relatively minor role in regulating directional growth in higher plants, although they can be important in certain specialized situations.

Considerable chemical interaction between plants does occur, of course. Many bacteria and fungi release hormones and hormone-like materials into the soil and these can have significant effects on root growth (references in Bilderback 1985). Again, in the phenomenon of allelopathy, higher plants themselves release into both the aerial and soil environments a very wide range of chemicals that can have strong inhibitory effects on the germination and growth of other plants (see Whittaker 1975, Bilderback 1985). However, in all these situations there are no reports of directional regulation of growth.

In some older physiological texts, it is suggested that roots can show chemotropic responses to soil nutrients. These suggestions seem generally to arise from a study in which chemicals were unilaterally applied to the individual roots of several species by various methods involving solutions, drips, and wicks (Newcombe & Rhodes 1904). Of course, such methods do not constitute a truly chemotropic directional assay (repositioning of stimulus? specificity of effect?), and in any case the results showed that only in one case, *Lupinus albus*, was there a slight response to a chemical, a solution of dibasic sodium phosphate. Again, it has been reported that certain parasitic plants, such as dodder (*Cuscuta* sp.), are chemotropically attracted towards their hosts, and that the haustoria may even be chemotropically attracted towards particular nutrient-rich tissues, such as the phloem (see Ziegler 1962a), but these phenomena have been little investigated.

Certain plant organs, however, are generally believed to show directional responses to chemical stimuli. These include the trapping organs of insectivorous plants, and the pollen tubes of plants in general during their growth down the style. But even in these traditional examples of chemotropism in higher plants, the questions to bear in mind are:

(a) Is there a directional response to an actual gradient of a chemical?

(b) Is there a specific response, which is likely to involve the action of a specific chemoreceptor?

Insectivorous plants

Mechanical stimuli generally trigger the initial release of insect traps in plants (see Section 5.3), but in many cases chemical stimuli seem to be responsible for the subsequent tightening of the trap (Ziegler 1962a). For example, in the Venus flytrap (Fig. 5.2), the two lobes of the trap continue to tighten on the insect for about 10 days; in sundew (Fig. 5.3) the tentacles that have not received any mechanical stimulation eventually also close over the trapped insect; and in butterwort (*Pinguicula* sp.) the insect is initially trapped in the sticky mucilage on the leaf surface, then the edges of the leaf curve inwards over the insect during the subsequent few hours.

All of these later movements, presumably growth responses, seem to be induced by chemicals. However, there is little specificity in any of the responses, low molecular weight nitrogenous compounds and sodium being generally effective (Lichtner & Williams 1977). In sundew, the head cells of the tentacles seem to be chemosensitive as well as thigmosensitive (Aletsee 1962b), but nothing is known about any chemoreceptors.

The pollen tube

There are very many compatibility and recognition reactions between pollen and stigmatoid tissues (see Knox 1984). However, since the early work of Molisch and of DuBary in the 1890s, the subsequent directional growth of the pollen tube down the style and into the ovary has generally been considered to be the classic example of chemotropic growth in higher plants. Certainly, the pollen tube must make several substantial changes of direction in penetrating from the style first to the ovary, then to an ovule, and eventually to the embryo sac of the ovule. However, it is unclear exactly which stylar chemicals may be involved in regulating the direction of growth of the pollen tube and, further, whether any such regulation involves response to an actual gradient of chemicals.

The stylar tissues of flowers can be categorized into three general types according to the means by which they accommodate growth of the pollen tube (Knox 1984):

(a) 'Open' styles, in which the pollen tubes grow down through a stylar canal that is filled with mucilage secreted from the epithelial cells of the canal; this type of style is characteristic of the monocots, particularly the Liliaceae.
(b) 'Closed' styles, in which the pollen tubes grow down through the stylar tissues; this is characteristic of most dicots, although some have open styles.

(a)   (b)

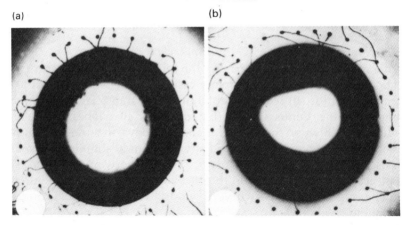

Figure 6.4 Effects of stigmatoid and stylar tissues on the *in vitro* germination and growth of pollen tubes of *Lilium longiflorum*: (a) an aqueous extract of stigma and style tissues was placed in the centre well of the agar; (b) distilled water in the centre well (from Rosen 1961).

(c) 'Semi-closed' styles, which exhibit intermediate features and which are characteristic of the Cactaceae.

Studies of chemotropic responses in pollen tubes have been largely carried out on species of the Liliaceae. An early study in *L. auratum* concluded that pollen tubes were only attracted towards stigmatoid tissues and, on the basis of inversion experiments, that their subsequent growth down the style was directed by gravity (Iwanami 1953). However, later studies suggested that the cells of both the stigma and the stylar canal secreted substances which had a chemotropic influence on tube growth (Rosen 1971). The nature of these substances is problematical.

Investigations in this area have generally relied simply upon observing the behaviour of pollen germ tubes (i.e. early stages of growth) growing on agar in proximity to pieces of stigma or style (Fig. 6.4). It has been pointed out that the general lack of re-positioning of the stimulus and the often limited periods of growth, result in these being not very strict tests of chemotropism (van den Ende 1976). Nevertheless, on the basis of such *in vitro* studies there are suggestions that the chemotropically effective components in the mucilage are sugars and amino acids. (Alternatively, the observed growth responses could be due to the 'straightening effects' of growth stimulation.) Other studies have demonstrated directional effects of calcium on tube growth. However, stigmatoid canal cells are actually lower in calcium than the surrounding cells and, further, no calcium gradient has been detected along the *in vivo* growth path of the pollen tube (Glenk *et al.* 1971).

Present opinion regarding the directional control of pollen tube growth (in open styles) is that the canal cells probably secrete nutrients or precursors which are directly incorporated into the apical growing point of the tube (van den Ende 1976). It has also been suggested that some factor in the canal mucilage may weaken or loosen the pollen tube wall at the tip, and that the turgor pressure within the pollen tube then causes the wall to bulge, and grow, in that direction (Rosen 1961). In either of these situations, it is a moot point whether growth of the pollen tube is strictly chemotropic, in the sense of being directional along the gradient of a chemical and involving specific chemoreceptors.

Box 6.1   Temperature effects and thermosensing

Temperature exerts marked effects, of course, on all aspects of plant growth and development through its general action on metabolic and physiological processes, and this field of 'thermophysiology' continues to receive considerable attention. The question being posed here, however, concerns the possibility that some of these responses may represent specific processes of 'thermosensing' in plants, involving specific receptors of thermal energy whose action initiates specific, sensory transduction sequences.

In animals, specialized nerve cells act as thermoreceptors, which are classed as 'warm' or 'cold' receptors according to whether they discharge as a result of a rise or a fall in temperature. (In some animals, such receptors are very highly developed, e.g. the thermoreceptors in the facial pits of rattlesnakes can discharge in response to a discrete temperature rise of 0.002 °C.) The extent to which thermosensing also occurs in aneural organisms has been considered in a significant review by Poff *et al.* (1984), and their conclusions may have a bearing on some plant responses.

These authors first of all consider the characteristics of certain temperature responses in several organisms. They note that both *Escherichia coli* and *Paramecium* show 'temperature adaptation', i.e. the temperature at which the organisms have been cultured determines the position at which they subsequently accumulate in temperature gradients. Furthermore, sudden changes in temperature can affect the frequency of tumbling in *E. coli* (cf. bacterial chemotaxis, Section 6.1) and can induce transient reversal of ciliary beat in *Paramecium* (cf. responses to mechanostimuli, Section 5.3). Again, the cellular slime mould *Dictyostelium* is very sensitive to temperature and can orient itself in a gradient of 0.05 °C cm^{-1} (representing a temperature differential of 0.0005 °C across the cell); it also shows 'adaptation' in that the sensitivity range is influenced by the temperature at which the organism has been cultured. Nematodes also accumulate in a temperature gradient at a point that is determined by the culture temperature; and in these organisms thermotactically defective mutants have been observed.

The authors conclude that all these characteristics of adaptation, extreme sensitivity, and mutability indicate specific thermosensing, rather than general temperature effects on metabolism. They suggest that, by analogy

175

with other sensory processes, the 'thermoreceptors' are likely to be associated with the cell membrane, and further, that they may involve the membrane lipids. In support of this latter suggestion they point out that temperature not only affects the physical properties of lipids (and thus, possibly the ionic permeability of the membrane), but it also affects the lipid composition of the membrane (and thus possibly provides a basis for temperature adaptation).

Since animal cells, microbial cells, and protozoan cells all seem to have specific thermosensing systems, it may be that plant cells also have a similar sort of capability. In plants there is no convenient 'locomotory assay' with which to investigate characteristics of adaptation, sensitivity, and mutability. However, observed thermo-induced events in plants that may be related to some form of thermosensing include the generation of action potentials in response to changes in temperature (Davies 1987, Edwards & Pickard 1987), and changes in membrane lipid composition as a result of changes in growth temperature (see Öquist 1983).

6.2 HYDROTROPISM

6.2.1 Early background

The notion that plant roots penetrate the soil in search of water to sustain their growth may be one of those popular ideas that persist, simply because it is so easily thought of and seems so natural. Indeed, this sort of behaviour, rather than the action of gravity, was first offered as the explanation for the downward orientation of roots (suggested by Dodart, around 1700). However, with greater awareness of the roles of other directional signals such as gravity and light on the general orientation of plant organs, interest in hydrotropism, and belief in its reality, has fluctuated over the years.

The agriculturalist T. A. Knight (whose earlier waterwheel experiments had related centrifugal force to gravitropism, see Section 2.2) was one of the first to specifically observe that water can influence root orientation (from Hooker 1915). In 1811, Knight reported that roots grew out of the bottoms of flower pots into air or soil only if the air or external soil was saturated with moisture; if the external medium was dry, the roots remained adpressed along the undersurface of the pot. By 1872 Sachs had placed this observation on a more experimental basis (Fig. 6.5); he demonstrated that in seedlings grown at an angle in a sieve basket, the emergent roots became diverted from the vertical and grew along the bottom of the basket. Some investigators were unable to repeat this demonstration (because, Sachs suggested, they had not taken due account of other factors such as light, temperature and moisture content of the surrounding air). Nevertheless, around that time, Sachs, Darwin,

Figure 6.5 Hydrotropism in roots, as demonstrated by the 'hanging basket' technique of Sachs; when the roots emerge from the bottom of the basket they become diverted from the vertical (from Ziegler 1962b, after Sachs).

Pfeffer, and Weisner (who introduced the term hydrotropism) were all convinced that moisture affected root orientation.

Subsequently, however, there have been two quite different types of opinion regarding the extent and the significance of the phenomenon. These can perhaps be best illustrated by describing two different types of approach that have been used to investigate it.

6.2.2 Vapour gradients

The first type of approach is exemplified by the studies of Hooker (1915) who investigated the behaviour of roots grown in air containing various gradients of water. Seedlings of *Lupinus albus* were grown in closed tanks in particular moisture vapour gradients that had been created by placing the seedlings at appropriate distances between a 'wetting agent' of soaked paper and a 'drying agent' of sulphuric acid. The actual moisture gradients were measured, and the behaviour of the roots observed accordingly. Hooker found that the roots of this species were 'positively hydrotropic', with curvature apparent within 6 hours, but also that the conditions for obtaining the response were extremely critical. The response was only obtained with a vapour content of 80–100% relative moisture (below 80% moisture the roots stopped growing). Furthermore, the minimum moisture gradient to which the roots reacted was a fall of 0.2% cm^{-1} (bending 10° from the vertical); the optimum response was obtained in a gradient of 0.4% cm^{-1} (60° from the vertical); and no response was obtained if the gradient was steeper than 0.5% cm^{-1} (i.e. in steep gradients the 'dry' side of the root simply dried out, causing a physical bending away from moisture). These physical effects of the root drying unilaterally in the steep gradients tend to highlight the rather surprising aspect of the positive physiological curvature in the shallower gradients, i.e. it was the 'drier' side of the organ that was elongating more rapidly. Hooker also ruled out experimentally that the growth effects were due to temperature differences across the root, caused by differential evaporation. (In fact, in this respect he was of the opinion that earlier reports of 'thermotropism' were really due to evaporational water loss inhibiting growth on the warmer side.)

These studies therefore seem to validate the occurrence of directional growth responses in relation to moisture. However, they say nothing about possible mechanisms, tropic or otherwise. And they indicate the extremely critical nature of the conditions that are necessary for their expression.

6.2.3   Field trials

Loomis & Ewan (1936) studied the possible occurrence of root hydrotropism under more natural growing conditions. They created a moisture gradient by making use of the fact that at a certain moisture content (termed by these authors the 'field coefficient') a soil does not lose moisture by capillary action to dry soil that is in contact with it; at the field coefficient, however, there is still more than enough moisture freely available to support plant growth. In the case of the soils used by Loomis and Ewan the relevant moisture contents were:

$$\text{field coefficient} = 13.8\% \text{ moisture}$$
$$\text{optimum for plant growth} = 11.8\% \text{ moisture}$$
$$\text{plant wilting coefficient} = 7.1\% \text{ moisture}$$

A layer of soil at a moisture content slightly below its field coefficient ('wet soil', 11.8%) was placed in contact with soil at a moisture content below its hygroscopic coefficient ('dry soil', 4.8%). Seedlings of various species were grown in the wet soil, and observations made on the directional responses of the roots when they grew near or across the 'moisture line'.

Results varied with the species. In most, the roots stopped growing when they entered the dry soil and branching was stimulated in the part of the root remaining in the wet soil; when any of these branch roots entered the dry soil, again their growth stopped and further branching was stimulated in the remainder of the root. Thus, there was no truly directional growth response towards moisture, but greater root development in the moist conditions gave a directional result (more or less by a form of nutritional entrapment).

The roots of some species showed signs of directional responses towards the moist conditions. For example, when the moisture line was oriented at an angle of 45°, these roots diverted from their downward path and followed the moisture line; and in a few species the roots followed the moisture line even when it was oriented horizontally. (However, it may be that the moist conditions offered less impedance and that this behaviour actually represented thigmotropic responses, see Section 5.2.)

Perhaps, therefore, the roots of some plants can show directional responses to moisture under particular conditions. However, in plants in general, and particularly under the range of moisture conditions found in nature, the effects of moisture on the developmental and branching patterns of the root are probably of much greater significance. Furthermore, in the limited number of species in which a moisture gradient may induce differential growth, it is difficult to conceive of this as being due to anything other than non-specific effects on the general physiology of the root. That is, there is unlikely to be a specific 'water receptor' which initiates a specific directional growth response.

Box 6.2   Directional responses to electrical stimuli

Electrical activity in plants was first detected over 100 years ago. Phenomena now studied in this area include:

(a)   The bioelectric potentials that exist across cell membrances; these are of the order of 10–200 mV, which correspond to field strengths across a 10 nm membrane of 10^3–10^4 V cm^{-1} (Fensom 1985).

(b)   The internal and external loops of current that occur in growing zones of organs (Bentrup 1984); these are associated with ion fluxes and consist of current densities of around 10 mA m^{-2}, i.e. equivalent to about 100 μmol m^{-2} s^{-1} (Miller et al. 1986).

(c)   The electrical signals that are transmitted throughout plant tissues at rates of 0.2–200 mm s^{-1} (and even faster in some specialized organs); these often consist of action potentials, which can be generated in response to a very wide range of external stimuli (Pickard 1973, Bentrup 1979, Davies 1987).

However, the question here is whether plants detect and respond to external electric stimuli.

When an electrical potential difference at fairly high voltage is applied across plant tissues, various things can happen (Fensom 1985): at high voltages, the tissues may produce heat and burn; at lower voltages (<50 V cm^{-1}), damage may result from ruptured membranes (giving wound responses rather than electrical effects); at low voltages (<1.5 V cm^{-1}) ionic disturbances may occur and polar molecules may become reoriented.

In higher plants, electrical stimulation can induce responses in specialized tissues such as those of *Mimosa* (Bentrup 1979) and some tendrils (Jaffe 1985). However, in general, growth and developmental responses to applied electrical fields or currents have been described as usually inhibitory and often inconsistent (Fensom 1985). Electro-induced directional responses are in an even greater state of confusion, with a great many contradictory reports regarding types of response and directions of curvature (Schrank 1959). In most of this early work, usually on isolated or decapitated organs, responses were obviously confounded by indirect effects of cellular damage and electrolytic by-products. Overall, no clear picture of responses to applied electric stimuli has emerged. It may be that their lack of a sophisticated electrical system for the transmission of signals means that higher plants can only respond in the crudest way to relatively massive inputs of external electrical stimulation.

The situation may be different in unicellular organs and lower plants. It has been suggested, for example, that the extracellular ion currents that exist around the growing zones of roots may serve as guidance signals for the zoospores and hyphae of root pathogens, thus ensuring that not only can a root be located (say by chemotaxis towards exudate) but that it is a living root (Miller *et al.* 1986, 1988). In this respect it has been pointed out (Carlile 1980) that there is a distinction between response to an electric field (electrotaxis, electrotropism) and response to an electric current (galvanotaxis, galvanotropism).

6.3   TRAUMATROPISM

Wounding is a mechanical process which ruptures or damages cells in a localized region, but which also generally results in changes in the activities of cells in other regions (Imaseki 1985). These changes in activity include the stimulation of growth and cell division in previously quiescent cells, the reconnection of vascular elements, and the eventual initiation of new organs, in a sequence of events that lasts several days or even weeks. However, wounding can also often bring about differential growth in the wounded organ within the first hour or so. The term traumatropism (or traumatotropism) was introduced by Pfeffer in 1893 to describe such wound-induced, directional growth responses. [Different types, and extents, of wounding may bring about quite different forms of metabolic and physiological responses. For example, PIIFs ('proteinase inhibitor inducing factors', see later) are formed in leaf cells in response to crushing, but seemingly not in response to slicing (Davies 1987). However, in the literature on traumatropism, no serious attempt seems to have been made to distinguish between the effects of different types of wounding.]

6.3.1 Traumatropic growth responses

In conjunction with his experiments on thigmotropism (see Section 5.2) Darwin (1880) also investigated the responses of roots to actual injury. In general, he found much the same form of response to wounding as to touch, i.e. if the wound was close to the root tip, the root curved away from the side that was wounded, occasionally even to the extent of forming a 90° bend; but if the wound was slightly farther back from the tip, the root curved towards the wounded side. The American investigator Spalding (1894) confirmed these findings, but considered that the positive curvature towards the sub-apically wounded side was simply due to mechanical damage to the elongation zone. Therefore, it is generally accepted that roots show negative traumatropic curvature, and that this curvature is due to differential growth in a region distinct from the site of the wound. However, the nature of the differential responses, whether by differential growth stimulation or inhibition, is not known.

Aerial organs, in general, seem to be much less sensitive than roots to wound stimuli. Most work has been carried out on the coleoptile, in which curvatures of 30°–40° can be induced by slicing or abrasion. Early studies indicated that wounding induced positive curvature towards the affected side. Subsequently, however, it was noted that after this initial curvature, the organ curved back to its original orientation (see Bünning 1959b). Therefore by analogy with roots, this 'negative curvature' was for a time considered to be the true traumatropic response. (The 'negative' curvature was not simply gravitropic recovery, since it also occurred on a klinostat.) Further studies indicated yet a third phase of response in which the organ curved back towards the wounded side. The extents, timings, and even general occurrence of these 'first positive', 'first negative', and 'second positive' traumatropic curvatures vary among different investigators.

Consideration of the actual growth responses that can be induced by wounding may clarify some of this behaviour (see Bünning 1959b). The initial response, within minutes, consists of a strong inhibition of growth around the site of the wound, giving the initial rapid positive curvature. However, a considerable amount of growth stimulation closely follows, or sometimes even occurs together with, the inhibition. This growth stimulation is distributed more widely than the localized inhibition, and is particularly evident on the unwounded side of the organ, thus eventually causing a positive curvature towards the wounded side. It is the extent and timing of the growth stimulation that seems to vary somewhat between different traumatropic treatments.

Therefore, in their negative and positive curvatures, roots and shoots show opposite forms of traumatropic response (behaviour that may be of adaptive significance in their respective subterranean and aerial environments). However, in both organs the directional changes derive from growth responses in regions that are distinct from the site of wounding, indicating the involvement of some kind of message transmission. The nature of this message is discussed in the next section.

6.3.2 Possible mechanisms of regulation

A large amount of research has been carried out on the metabolic responses of wounded tissues, particularly storage tissues (see Imaseki 1985, Davies 1987). Aspects studied include changes in respiration rate, in the syntheses of proteins and nucleic acids, and in enzyme activities. However, the relevance of these kinds of biochemical changes in wounded tissues to the initiation of differential growth in cells some distance from the site of wounding is not yet clear.

There are three different, though possibly interrelated, means by which wounding may result in the transmission of growth regulatory messages in plant tissues: chemically (through the movement and action of hormonal factors), electrically (through the generation and propogation of action potentials), and perhaps also mechanically (through readjustment in the planes of mechanical stress within an organ).

Chemical regulation
In the early literature there were suggestions that wounding affected growth through the loss or deprivation of chemical growth factors (see Bünning 1959b). However, in aerial organs directional effects on growth can be induced simply by abrasion of the epidermis, and do not require complete disruption of the vascular or any other tissue; moreover, the responses also involve growth stimulation, not just inhibition. Therefore it is now generally thought that these responses are likely to result from the induction or the presence of a regulator, rather than from the absence of something. (The negative traumatropic curvature of roots, however, could result from a wound-induced disruption in the supply of growth inhibitor on one side of the root cap.)

A common response of plant organs to wounding (Imaseki 1985) and other stressful stimuli (see Kang 1979, Palmer 1985, Pickard 1985a) is an enhanced production of ethylene. The kinetics of the ethylene wound response vary among different organs, but there is generally a lag period of 20–30 minutes, after which ethylene is produced at high levels, continually or with some fluctuations, for periods of hours (e.g. etiolated pea stem segments) or days (e.g. bruised apples). Ethylene, of course, is usually associated with effects on growth inhibition. This, together with its production by cells that are adjacent to the site of a wound, suggests that it may be involved in the initial, localized inhibition of growth in traumatropism, but its kinetics of continued production do not correlate well with the subsequent rapid changes in growth pattern.

In *Mimosa*, the slow-moving electrical potential that is induced by wounding (see Section 5.3.1) may be associated with some kind of chemical 'wound factor' (Ricca's factor) that is released into the xylem (Roblin 1979). This factor is thought to be carried along in the transpiration stream and to be responsible for triggering changes in the electrical activities of neighbouring cells (and eventual turgor changes in the motor cells of pulvini). The nature of this particular chemical is unknown, but it has recently been equated with the general class of turgorins (Schild-knecht 1984), i.e. derivatives of certain organic acids that are involved in turgor-based leaf movements (see Section 1.2.2). However, there is as yet no known relationship between wound-induced turgorins and the growth responses of wounded tissues.

In 1917, Haberlandt suggested that a specific wound hormone was produced in response to injury (see Imaseki 1985). From time to time substances that seem wound-specific have been isolated, but these have invariably been found not to have general regulatory effects on growth (see Bünning 1959b). For example, traumatic acid seems to appear as a specific response to wounding in beans, but it had no effects on growth when it was applied back to unwounded organs (Galston & Davies 1970).

Recent research into the biochemistry of plant defence mechanisms has indicated that cell wall fragments can act as powerful signalling agents, not only to activate defence genes but also to initiate profound changes in growth and development (Ryan 1987). These wall fragments have received the general names of 'proteinase inhibitor inducing fragments' (PIIFs) and 'oligosaccharins', and they seem to be produced in response to a wide variety of wound effects. However, they are unlikely to be the vehicles of traumatropic message transmission: in the first place these materials do not themselves seem to be transported around the plant, but rather act as secondary messengers in a cell in response to some other transmitted signal (Davies 1987); and secondly they do not seem to be induced in response to wounding by razor-cutting (Davies 1987), a stimulus that is quite effective in inducing a traumatropic response.

Other (non-tropic) situations of wounding can also bring about changes in growth rate in regions some distance from the site of the wound, and in some of these cases the characteristics of the growth responses are not consistent with message transmission by means of a hormone. For example, in oat coleoptiles subjected to complete decapitation or only partial apical wounding, the resultant growth inhibition does not appear in a sequential pattern along the coleoptile from the apex to the base, behaviour inconsistent with the inhibitory effects being due to (or 'transmitted' by) diminution in auxin movement from the apex

(Parsons *et al.* 1988); indeed within 30 minutes of wounding, an effect on growth is apparent in a region 15 mm distant, a rate of stimulus transmission that is inconsistent with the accepted rates of auxin movement (10 mm h^{-1}), and perhaps also with the rate of movement of any regulatory chemical.

Electrical (ionic) regulation

Mechanisms to account for the general effects of wounding other than through deprivation or disruption of the hormonal supply have been proposed. One such model suggests that effects result from damage to the membrane, particularly to the electrogenic ion pump, with consequent membrane depolarization and disruption of ion transport (Hanson & Trewavas 1982).

In this respect, wounding can induce measurable changes in membrane potential (Bentrup 1979). For example in etiolated peas, wound-induced action potentials show peaks lasting a minute or so and travelling through the tissues at rates of 3–5 cm min^{-1} (Davies 1987). Such action potentials are often associated with turgor-based responses, not only in specialized tissues, e.g. in *Mimosa* (see Section 5.3), but also in other organs, e.g. in leaves wounding has been observed to generate a slow-moving electrical signal that seems to be correlated with partial stomatal closure (Simons 1981). In other situations, the generation of action potentials seems to be associated with responses that also involve changes in growth rates, e.g. tendril coiling (Pickard 1973, Jaffe 1985). And in seedlings of *Bidens pilosus*, pricking a cotyledon results in the transmission of a wave of electrical depolarization that seems to correlate with the subsequent inhibition of growth of the bud in the axil of that cotyledon (Frachisse *et al.* 1985).

Action potentials represent regenerative changes in the properties of the cell membrane, and in plants seem to be based generally on potassium efflux from the cell and calcium influx (see Section 2.1). Recently a unifying hypothesis has been put forward to account for the regulatory effects of action potentials (Davies 1987). This proposes that not only do action potentials serve as a major means of long-distance communication in plants, but also that the various components of an action potential, separately or together, act to regulate cellular activities at the local level. For example, it is suggested that the potassium (efflux) component may be specifically involved in initiating cell turgor responses, whereas the calcium (influx) component may be involved in the regulation of other cellular activities, including processes of growth. Whether or not this latter aspect, calcium regulation of growth responses, is involved in traumatropism remains to be discovered.

Mechanical regulation

It is well established that the planes of cell division in a plant are generally related to the planes of mechanical stress throughout the organ, and it has been demonstrated that if these stress planes are altered then the planes of cell division alter accordingly (Lintilhac & Vesecky 1981, 1984). Cell division represents a relatively long-term growth response. However, as discussed in Section 5.3.2, if there is any form of 'stretch receptor' in the plant cell membrane (Edwards & Pickard 1987), then mechanical stimuli such as changes in tension or compression are likely also to affect rates of elongation in the shorter term, either directly or through changes in the sensitivity of the cells to growth regulators.

It has also been pointed out in Section 5.3.2 that a growing plant organ represents a mechanical equilibrium of mutually opposing forces. Any disturbance to this equilibrium must result in the creation of new planes of mechanical stress. Since the epidermis seems to act as a regulatory compressive force on the internal tissues (see Box 3.2), then even superficial damage to this tissue must give rise to changes in the stress pattern, and thus perhaps to mechanically stimulated changes in cellular growth rates. Such events may be involved in directional growth responses to wounding.

However, uncertainty about the actual regulatory basis of traumatropism does not detract from the general significance of plant growth responses to wounding and injury. They are of significance, of course, to the investigative physiologist who should be aware (or beware) that any wounding of a growing organ generally results in rapid and substantial fluctuations in growth rate in different regions of the organ over a considerable number of hours. But they are also of significance to the plant itself, with directed growth movements again playing a role, this time protectively, in the general responses of the plant to its environment.

References

Aletsee, L. 1962a. Thermotropismus. In *Physiology of movements*, Encyclopaedia of Plant Physiology, Vol. 17, part 2, W. Ruhland (ed.), 1–14. Berlin: Springer.

Aletsee, L. 1962b. Chemonastie. In *Physiology of movements*, Encyclopaedia of Plant Physiology, Vol. 17, part 2, W. Ruhland (ed.), 451–83. Berlin: Springer.

Arslan-Cerim, N. 1966. The redistribution of radioactivity in geotropically stimulated hypocotyls of *Helianthus annuus* pretreated with radioactive calcium. *Journal of Experimental Botany* **17**, 236–40.

Audus, L. J. 1969. Geotropism. In *Physiology of plant growth and development*, M. B. Wilkins (ed.), 205–42. London: McGraw-Hill.

Audus, L. J. 1979. Plant geosensors. *Journal of Experimental Botany* **30**, 1051–73.

Avery, G. S. & P. R. Burkholder 1936. Polarised growth and cell studies on the *Avena* coleoptile test object. *Bulletin of the Torrey Botanical Club* **63**, 1–15.

Badham, E. R. 1984. Measuring curvature in cylindrical plant organs. *Experimental Mycology* **8**, 176–8.

Baillaud, L. 1959. Anatomie physiologique des organes thigmotropiques et thigmonastiques. In *Physiology of movements*, Encyclopaedia of Plant Physiology, Vol. 17, part 1, W. Ruhland (ed.), 243–53. Berlin: Springer.

Ball, N. G. 1962. The effect of externally applied IAA on phototropic induction and response in coleoptiles of *Avena*. *Journal of Experimental Botany* **13**, 45–60.

Balmer, R. T. & J. G. Franks 1975. Contractile characteristics of *Mimosa pudica*. *Plant Physiology* **56**, 464–7.

Banbury, G. H. 1959. Phototropism of lower plants. In *Physiology of movements*, Encyclopaedia of Plant Physiology, Vol. 17, part 1, W. Ruhland (ed.), 530–78. Berlin: Springer.

Banbury, G. H. 1962. Geotropism of lower plants. In *Physiology of movements*, Encyclopaedia of Plant Physiology, Vol. 17, part 2, W. Ruhland (ed.), 344–77. Berlin: Springer.

Bandurski, R. S., A. Schulze, P. Dayanandan & P. B. Kaufman 1984. Response to gravity by *Zea mays* seedlings. I Time course of the response. *Plant Physiology* **74**, 284–8.

Barlow, P. W. & E. L. Rathfelder 1985. Distribution and redistribution of extension growth along vertical and horizontal gravireacting maize roots. *Planta* **165**, 134–41.

Baskin, T. I. 1986. Redistribution of growth during phototropism and nutation in the pea epicotyl. *Planta* **169**, 406–14.

Baskin, T. I. & M. Iino 1987. An action spectrum in the blue and ultraviolet for phototropism in alfalfa. *Photochemistry & Photobiology* **46**, 127–36.

Baskin, T. I., W. R. Briggs & M. Iino 1986. Can lateral redistribution of auxin account for phototropism of maize coleoptiles? *Plant Physiology* **81**, 306–9.

Baskin, T. I., M. Iino, P. A. Green & W. R. Briggs 1985. Measurements of growth during first positive blue-light induced phototropism in corn. *Plant, Cell & Environment* **8**, 595–603.

Bean, B. 1984. Microbial geotaxis. In *Membranes and sensory transduction*, G. Colombetti & F. Lenci (eds), 163–98. New York: Plenum Press.

Behrens, H. M., D. Gradman & A. Sievers 1985. Membrane potential responses following gravi-stimulation in roots of *Lepidium sativum*. *Planta* **163**, 463–72.

Behrens, H. M., M. H. Weisenseel & A. Sievers 1982. Rapid changes in the pattern of electric current around the root tip of *Lepidium sativum* following gravistimulation. *Plant Physiology* **70**, 1079–82.

Bentrup, F. W. 1979. Reception and transduction of electrical and mechanical stimuli. In *Physiology of movements*, Encyclopaedia of Plant Physiology, NS Vol. 7, W. Haupt & M. E. Feinleib (eds), 42–70. Berlin: Springer.

Bentrup, F. W. 1984. Cellular polarity. In *Cellular interactions*, Encyclopaedia of Plant Physiology, NS Vol. 17, H. F. Linskins & J. Heslop-Harrison (eds), 472–90. Berlin: Springer.

Berg, A. R., I. R. MacDonald, J. W. Hart & D. C. Gordon 1986. Relative elemental elongation rates in the etiolated hypocotyl of sunflower: a comparison of straight growth and gravitropic growth. *Botanical Gazette* **147**, 373–82.

Bilderback, D. E. 1984. Phototropism of *Selaginella*: the differential response to light. *American Journal of Botany* **71**, 1323–37.

Bilderback, D. E. 1985. Regulators of plant reproduction, growth and differentiation in the environment. In *Hormonal regulation of development, III Role of environmental factors*, Encyclopaedia of Plant Physiology, NS Vol. 11, R. P. Pharis & D. M. Reid (eds), 652–706. Berlin: Springer.

Bjorkman, T. 1988. Perception of gravity by plants. *Advances in Botanical Research* **15**, 1–41.

Bjorkman, T. & R. E. Cleland 1988. The role of the epidermis and cortex in gravitropic curvature of maize roots. *Planta* **176**, 513–18.

Bjorkman, T. & A. C. Leopold 1987a. An electric current associated with gravity sensing in maize roots. *Plant Physiology* **84**, 841–6.

Bjorkman, T. & A. C. Leopold 1987b. Effects of inhibitors of auxin transport and of calmodulin on a gravisensing-dependent current in maize roots. *Plant Physiology* **84**, 847–50.

Blaauw, O. H. & G. Blaauw-Jansen 1970. The phototropic responses of *Avena* coleoptiles. *Acta Botanica Neerlandica* **19**, 755–63.

Blatt, M. R. 1987. Toward the link between membranes, transport and photoreception in plants. *Photochemistry & Photobiology* **45**, 933–8.

Brennan, T., J. E. Gunckel & C. Frenckel 1976. Stem sensitivity and ethylene involvement in phototropism of mung bean. *Plant Physiology* **57**, 286–9.

Bridges, I. G. & M. B. Wilkins 1973. Growth initiation in the geotropic response of the wheat node. *Planta* **112**, 191–200.

Briggs, W. R. 1963. The phototropic responses of higher plants. *Annual Review of Plant Physiology* **14**, 311–52.

Briggs, W. R. & M. Iino 1983. Blue light absorbing photoreceptors in plants. *Philosophical Transactions of the Royal Society of London* **B303**, 347–57.

Briggs, W. R., R. D. Tocher & J. F. Wilson 1957. Phototropic auxin redistribution in corn coleoptiles. *Science* **126**, 210–12.

Britz, S. J. & A. W. Galston 1982. Physiology of movements in stems of seedlings of *Pisum sativum* cv. Alaska. I. Experimental separation of nutation from gravitropism. *Plant Physiology* **70**, 264–71.

Brown, A. H. & D. K. Chapman 1981. Comparative physiology of plant behaviour in simulated hypogravity. *Annals of Botany* **47**, 225–8.

Brown, A. H. & D. K. Chapman 1984. Circumnutation observed without a significant gravitational force in spaceflight. *Science* **225**, 230–2.

Brown, A. H. & D. K. Chapman 1988. Kinetics of suppression of circumnutation by clinostatting favours modified internal oscillator model. *American Journal of Botany* **75**, 1247–51.

Brummel, D. A. & J. L. Hall 1980. The role of the epidermis in auxin-induced and fusicoccin-induced elongation of *Pisum sativum* segments. *Planta* **150**, 371–9.

Bünning, E. 1959a. Die thigmonastischen und thigmotropischen reaktionen. In *Physiology of movements*, Encyclopaedia of Plant Physiology, Vol. 17, part 1, W. Ruhland (ed.), 254–77. Berlin: Springer.

Bünning, E. 1959b. Die wirkung von wundreizen. In *Physiology of movements*, Encyclopaedia of Plant Physiology, Vol. 17, part 1, W. Ruhland (ed.), 119–34. Berlin: Springer.

Burg, S. P. & E. A. Burg 1967. Lateral auxin transport in stems and roots. *Plant Physiology* **42**, 801–3.

Callow, J. A. 1984. Cellular and molecular recognition between higher plants and fungal pathogens. In *Cellular interactions*, Encyclopaedia of Plant Physiology, NS Vol. 17, H. F. Linskins & J. Heslop-Harrison (eds), 212–37. Berlin: Springer.

Carlile, M. J. 1965. The photobiology of fungi. *Annual Review of Plant Physiology* **16**, 175–202.

Carlile, M. J. 1975. *Primitive sensory and communication systems*. London: Academic Press.

Carlile, M. J. 1980. Sensory transduction in aneural organisms. In *Photoreception and sensory transduction in aneural organisms*, F. Lenci & G. Colombetti (eds), 1–22. New York: Plenum Press.

Caspar, T., C. Somerville & B. G. Pickard 1985. Geotropic roots and shoots of a starch-free mutant of *Arabidopsis*. *Plant Physiology* **77**, S–105.

Chance, H. L. & J. M. Smith 1946. The effects of light, gravity and centrifugal force upon the tropic responses of buckwheat seedlings. *Plant Physiology* **21**, 452–8.

Clifford, P. E. 1988. Early bending kinetics in response to geostimulation or unilateral IAA application in nodal segments. *Plant, Cell & Environment* **11**, 621–8.

Clifford, P. E. & G. F. Barclay 1980. The sedimentation of amyloplasts in living statocytes of the dandelion flower stalk. *Plant, Cell & Environment* **3**, 381–6.

Clifford, P. E., D. S. Fensom, B. I. Munt & W. D. McDowell 1982. Lateral stress initiates bending responses in dandelion peduncles: a clue to geotropism? *Canadian Journal of Botany* **60**, 2671–3.

Clifford, P. E., D. M. A. Monsdale, S. J. Lynd & E. L. Oxlade 1985. Differences in auxin level detected across geostimulated dandelion peduncles: evidence supporting a role for auxin in geotropism. *Annals of Botany* **55**, 293–6.

Coombe, D. & P. Bell (trans.) 1965. *Strasburger's textbook of botany*, New English edn. London: Longmans.

Cosgrove, D. J. 1981. Rapid suppression of growth by blue light. *Plant Physiology* **67**, 584–90.

Cosgrove, D. J. 1985. Kinetic separation of phototropism from blue light inhibition of stem elongation. *Photochemistry & Photobiology* **42**, 745–51.

Cosgrove, D. J. 1987. Wall relaxation and the driving forces for cell expansive growth. *Plant Physiology* **84**, 561–4.

Crombie, M. L. 1962. Thermonasty. In *Physiology of movements*, Encyclopaedia of Plant Physiology, Vol. 17, part 2, W. Ruhland (ed.), 15–33. Berlin: Springer.

Cumming, B. G. & E. Wagner 1968. Rhythmic processes in plants. *Annual Review of Plant Physiology* **19**, 381–416.

Curry, G. M. 1969. Phototropism. In *The physiology of plant growth and development*, M. B. Wilkins (ed.), 241–73. New York: McGraw-Hill.

Curry, G. M. & H. E. Gruen 1957. Negative phototropism of *Phycomyces* in the ultra-violet. *Nature* **179**, 1028–9.

Curry, G. M. & H. E. Gruen 1959. Action spectra for the positive and negative photoresponses of *Phycomyces* sporangiophores. *Proceedings of the National Academy of Sciences (USA)* **45**, 797–804.

Darwin, C. 1875a. *The movements and habits of climbing plants*. London: John Murray.

Darwin, C. 1875b. *Insectivorous plants*. London: John Murray.

Darwin, C. 1880. *The power of movement in plants*. London: John Murray.

Darwin, F. 1899. On geotropism and the localisation of the sensitive region. *Annals of Botany* **13**, 567–74.

Davies, E. 1987. Action potentials as multifunctional signals in plants: a unifying hypothesis to explain apparently disparate wound responses. *Plant, Cell & Environment* **10**, 623–31.

Davis, B. D. 1975. Bending growth in fern gametophyte protonema. *Plant and Cell Physiology* **16**, 537–41.

Daye, S., R. L. Biro & S. J. Roux 1984. Inhibition of gravitropism in oat coleoptiles by the calcium chelator, EGTA. *Physiologia Plantarum* **61**, 449–54.

Delbrück, M. & W. Shropshire 1960. Action and transmission spectra of *Phycomyces*. *Plant Physiology* **35**, 194–204.

Dennison, D. S. 1971. Gravity receptors in *Phycomyces*. In *Gravity and the organism*, S. A. Gordon & M. J. Cohen (eds), 65–72. Chicago: University of Chicago Press.

Dennison, D. S. 1979. Phototropism. In *Physiology of movements*, Encyclopaedia of Plant Physiology, NS Vol. 7, W. Haupt & M. E. Feinleib (eds), 507–66. Berlin: Springer.

Dennison, D. S. 1984. Phototropism. In *Advanced plant physiology*, M. B. Wilkins (ed.), 149–62. London: Pitman.

Dennison, D. S. & C. C. Roth 1967. *Phycomyces* sporangiophores: fungal stretch receptors. *Science* **156**, 1386.

Dickinson, S. 1977. Studies in the physiology of obligate parasitism. X Induction of responses to a thigmotropic stimulus. *Phytopathologische Zeitschrift* **89**, 97–115.

Digby, J. & R. D. Firn 1979. An analysis of the changes in growth rate occurring during the initial stages of geocurvature in shoots. *Plant, Cell & Environment* **2**, 145–8.

Eckert, R. & D. Randall 1983. *Animal physiology*, 2nd edn. San Francisco: W. H. Freeman.

Edwards, K. L. & B. G. Pickard 1987. Detection and transduction of physical stimuli in plants. In *Cell surface and signal transduction*, E. Wagner, H. Greppin & B. Millet (eds), 41–66. Berlin: Plenum Press.

Edwards, M. C. & D. J. F. Bowling 1986. The growth of rust germ tubes towards stomata in relation to pH gradients. *Physiological Molecular Plant Pathology* **29**, 185–96.

Ehleringer, J. & I. Forseth 1980. Solar tracking by plants. *Science* **210**, 1094–8.

Ellis, R. J. 1984. Kinetics and fluence response relationships of phototropism in the dicot *Fagopyrum esculentum* (buckwheat). *Plant and Cell Physiology* **25**, 1513–20.

Ellis, R. J. 1987. Comparison of fluence–response relationships of phototropism in light and dark grown buckwheat. *Plant Physiology* **85**, 689–92.

Ende, H. van den 1976. *Sexual interactions in plants.* London: Academic Press.

Ende, H. van den 1984. Sexual interactions in the lower filamentous fungi. In *Cellular interactions,* Encyclopaedia of Plant Physiology, NS Vol. 17, H. F. Linskins & J. Heslop-Harrison (eds), 331–49. Berlin: Springer.

Engelmann, W. & M. Schrempf 1980. Membrane models for circadian rhythms. *Photochemical and Photobiological Reviews,* Vol. 5, K. C. Smith (ed.), 49–86. New York: Plenum Press.

Erickson, R. O. & W. K. Silk 1980. The kinematics of plant growth. *Scientific American* **242**, 134–51.

Etzold, H. 1965. Der polarotropismus und phototropismus der chloronemen von *Dryopteris filix-mas. Planta* **64**, 254–80.

Evans, M. L., R. Moore & K.-H. Hasenstein 1986. How plants respond to gravity. *Scientific American* **255**, 100–7.

Everett, M. 1974. Dose response curves for radish seedling phototropism. *Plant Physiology* **54**, 222–5.

Fahn, A. 1982. *Plant anatomy,* 3rd edn. Oxford: Pergamon Press.

Feinleib, M. E. 1980. Photomotile responses in flagellates. In *Photoreception and sensory transduction in aneural organisms,* F. Lenci & G. Colombetti (eds), 45–68. New York: Plenum Press.

Feldman, L. J. 1984. Regulation of root development. *Annual Review of Plant Physiology* **35**, 223–42.

Fensom, D. S. 1985. Electrical and magnetic stimuli. In *Hormonal regulation of development. III Role of environmental factors,* Encyclopaedia of Plant Physiology, NS Vol. 11, R. P. Pharis & D. M. Reid (eds) 624–52. Berlin: Springer.

Filner, B. & R. Hertel 1970. Some aspects of geotropism in coleoptiles. *Planta* **94**, 333–54.

Findlay, G. P. 1984. Nastic movements. In *Advanced plant physiology,* M. B. Wilkins (ed.), 186–200. London: Pitman.

Firn, R. D. 1986a. Phototropism. In *Photomorphogenesis in plants,* R. E. Kendrick & G. H. M. Kronenberg (eds), 367–89. Dordrecht: Martinus Nijhoff.

Firn, R. D. 1986b. Growth substance sensitivity: the need for clearer ideas, precise terms and purposeful experiments. *Physiologia Plantarum* **67**, 267–72.

Firn, R. D. & J. Digby 1977. The role of the peripheral cell layers in the geotropic curvature of sunflower hypocotyls: a new model of shoot geotropism. *Australian Journal of Plant Physiology* **31**, 337–47.

Firn, R. D. & J. Digby 1979. A study of the autotropic straightening reaction of a shoot previously curved during geotropism. *Plant, Cell & Environment* **2**, 149–54.

Firn, R. D. & J. Digby 1980. The establishment of tropic curvatures in plants. *Annual Review of Plant Physiology* **31**, 131–48.

Firn, R. D. & A. B. Myers 1987. Hormones and plant tropisms. The degeneration of a model of hormonal control. In *Hormone action in plant development: a critical appraisal,* G. V. Hoad, J. R. Lenton, M. B. Jackson & R. K. Atkin (eds), 251–64. London: Butterworth.

Firn, R. D., J. Digby & A. Hall 1981. The role of the shoot apex in geotropism. *Plant, Cell & Environment* **4**, 125–9.

Fisher, J. B. 1985. Induction of reaction wood in *Terminalia* (Combretaceae); roles of gravity and stress. *Annals of Botany* **55**, 237–48.

Fortin, M. C., F. J. Pierce & K. L. Poff 1989. The pattern of secondary root formation in curving roots of *Arabidopsis thaliana. Plant, Cell & Environment* **12**, 337–40.

Frachisse, J.-M., M. O. Desbiez , P. Champagnat & M. Thellier 1985. Transmission of a traumatic signal via a wave of electric depolarisation and induction of correlations between the cotyledonary buds of *Bidens pilosus*. *Physiologia Plantarum* **64**, 48–52.

Franssen, J. M. & Bruinsma 1981. Relationships between xanthoxin, phototropism and elongation growth in the sunflower seedling *Helianthus annuus*. *Planta* **151**, 365–70.

Franssen, J. M., R. D. Firn & J. Digby 1982. The role of the apex in the phototropic curvature of *Avena* coleoptiles: positive curvature under conditions of continuous illumination. *Planta* **155**, 281–6.

Franssen, J. M., S. A. Cooke, J. Digby & R. D. Firn 1981. Measurements of differential growth causing phototropic curvature of coleoptiles and hypocotyls. *Zeitschrift für Pflanzenphysiologie* **103**, 207–16.

Friedrich, U. & R. Hertel 1973. Abhangigkeit der geotropischen krummung der *Chara*-rhizoide von der zentrifugalbeschleunigung. *Zeitschrift für Pflanzenphysiologie* **70**, 173–84.

Gaba, V. & M. Black 1983. The control of cell growth by light. In *Photomorphogenesis*, Encyclopaedia of Plant Physiology, NS Vol. 16A, W. Shropshire & H. Mohr (eds), 385–400. Berlin: Springer.

Galland, P. 1983. Action spectra of photogeotropic equilibrium in *Phycomyces* wild type and three behavioural mutants. *Photochemistry and Photobiology* **37**, 221–8.

Galland, P. & E. D. Lipson 1985a. Modified action spectra of photogeotropic equilibrium in *Phycomyces blakesleeanus* mutants with defects in genes *madA*, *madB*, *madC*, and *madH*. *Photochemistry and Photobiology* **41**, 331–5.

Galland, P. & E. D. Lipson 1985b. Action spectra for phototropic balance in *Phycomyces blakesleeanus*: Dependence on reference wavelength and intensity range. *Photochemistry and Photobiology* **41**, 323–9.

Galston, A. W. 1959. Phototropism of stems, roots and coleoptiles. In *Physiology of movements*, Encyclopaedia of Plant Physiology, Vol. 17, part 2, W. Ruhland (ed.), 493–529. Berlin: Springer.

Galston, A. W. 1983. Leaflet movements in *Samanea*. In *The biology of photoreception*, SEB Symposium No. 36, D. Cosens & D. Vince-Prue (eds), 541–59. Cambridge: Cambridge University Press.

Galston, A. W. & P. J. Davies 1970. *Control mechanisms in plant development*. New Jersey: Prentice Hall.

Gardner, G., S. Shaw & M. B. Wilkins 1974. IAA transport during the phototropic responses of intact *Zea* and *Avena* coleoptiles. *Planta* **121**, 237–51.

Glenk, H. O., W. Wagner & O. Schimmer 1971. Can calcium ions act as a chemotropic factor in *Oenothera* fertilisation? In *Pollen development and physiology*, J. Heslop-Harrison (ed.), 255–61. London: Butterworth.

Gooday, G. W. 1975. Chemotaxis and chemotropism in fungi and algae. In *Primitive sensory and communication systems*, M. J. Carlile (ed.), 155–204. New York: Academic Press.

Gordon, D. C., I. R. MacDonald & J. W. Hart 1982. Regional growth patterns in the hypocotyls of etiolated and green cress seedlings in light and darkness. *Plant, Cell & Environment* **5**, 347–53.

Gorer, R. 1968. *Climbing plants*. London: Studio Vista.

Goss, M. J. 1977. Effects of mechanical impedance on root growth in barley. I Effects on elongation and branching of seminal roots. *Journal of Experimental Botany* **28**, 96–111.

Goss, M. J. & R. S. Russell 1980. Effects of mechanical impedance on root growth in barley. III Observations on the mechanism of the response. *Journal of Experimental Botany* **31**, 577–88.

Goswami, K. K. A. & L. J. Audus 1976. Distribution of calcium, potassium and phosphorous in *Helianthus annuus* hypocotyls and *Zea mays* coleoptiles in relation to tropic stimuli and curvature. *Annals of Botany* **40**, 49–64.

Gradmann, H. 1925. Untersuchungen über geotropische Reizstoffe. *Jahrbuch wissenschaften Botanische* **64**, 201–48.

Greacen, E. L. 1986. Root response to soil mechanical properties. In *Transactions of the 13th Congress of the International Soil Science Society*, pp. 20–47. Amsterdam: Elsevier.

Guharay, F. & F. Sachs 1984. Stretch-activated single ion channel currents in tissue-cultured embryonic chick skeletal muscle. *Journal of Physiology* **352**, 685–701.

Haberlandt, G. 1914. *Physiological plant anatomy*. Translated from the 4th German edn by M. Drummond. London: MacMillan.

Hader, D.-P. 1979. Photomovement. In *Physiology of movements*, Encyclopaedia of Plant Physiology, NS Vol. 7, W. Haupt & M. E. Feinleib (eds), 268–309. Berlin: Springer.

Halevy, A. H. 1986. The induction of contractile roots in *Gladiolus grandiflorus*. *Planta* **167**, 94–100.

Halstead, T. W. & F. R. Dutcher 1987. Plants in space. *Annual Review of Plant Physiology* **38**, 317–45.

Hanson, J. B. & A. J. Trewavas 1982. Regulation of plant cell growth: the changing perspective. *New Phytologist* **90**, 1–18.

Harrison, M. A. & B. G. Pickard 1986a. Evaluation of ethylene as a mediator of gravitropism by tomato hypocotyls. *Plant Physiology* **80**, 592–5.

Harrison, M. A. & B. G. Pickard 1986b. Auxin asymmetry during 'wrong way' gravitropic curvature of tomato hypocotyls. *Plant Physiology* **80**, S–7.

Hart, J. W. 1988. *Light and plant growth*. London: Unwin Hyman.

Hart, J. W. & I. R. MacDonald 1981. Phototropism and geotropism in hypocotyls of cress. *Plant, Cell & Environment* **4**, 197–201.

Hart, J. W. & I. R. MacDonald 1984. Is there a role for the apex in shoot geotropism? *Plant Physiology* **74**, 272–7.

Hart, J. W., D. C. Gordon & I. R. MacDonald 1982. Analysis of growth during phototropic curvature of cress hypocotyls. *Plant, Cell & Environment* **5**, 361–6.

Hartmann, K. M., H. Menzel & H. Mohr 1965. Ein Beitrag zur Theorie der polarotropischen und phototropischen Krümmung. *Planta* **64**, 363–75.

Hasegawa, K., H. Noguchi, C. Tanone, S. Sando, M. Takada, M. Sakodas & T. Hashimoto 1987. Phototropism in hypocotyls of radish. IV Flank growth and lateral distribution of *cis*- and *trans*-raphanusanins in the first positive photo-tropic curvature. *Plant Physiology* **85**, 379–82.

Haupt, W. 1965. Perception of environmental stimuli orienting growth and movement in lower plants. *Annual Review of Plant Physiology* **16**, 267–90.

Haupt, W. 1983. The perception of light direction and orientation responses in chloroplasts. In *The biology of photoreception*, SEB Symposium No. 36, D. Cosens & D. Vince-Prue (eds), 423–42. Cambridge: Cambridge University Press.

Havis, A. L. 1940. Developmental studies with *Brassica* seedlings: relation of cell size and cell number to the hypocotyl length. *American Journal of Botany* **27**, 239–61.

Heathcote, D. G. 1981. The geotropic reaction and statolith movements following geostimulation of mung bean hypocotyls. *Plant, Cell & Environment* **4**, 131–40.

Heathcote, D. G. 1982. Nutation in georeacting roots. *Plant, Cell & Environment* **5**, 343–6.

Heathcote, D. G. & B. W. Bircher 1987. Enhancement of phototropic response to a range of light doses in *Triticum aestivum* coleoptiles in clinostat-simulated microgravity. *Planta* **170**, 249–56.

Hensel, W. 1984. A role of microtubules in the polarity of statocytes from roots of *Lepidium sativum*. *Planta* **162**, 404–14.

Hepler, P. K. & R. O. Wayne 1985. Calcium in plant development. *Annual Review of Plant Physiology* **36**, 397–439.

Hertel, R. 1971. Aspects of the geotropic stimulus in plants. In *Gravity and the organism*, S. A. Gordon & M. Cohen (eds), 40–50. Chicago: University of Chicago Press.

Hertel, R. 1980. Phototropism of lower plants. In *Photoreception and transduction in aneural organisms*, F. Lenci & G. Colombetti (eds), 89–105. New York: Plenum Press.

Hertel, R. 1983. The mechanism of auxin transport as a model for auxin action. *Zeitschrift für Pflanzenphysiologie* **112**, 53–67.

Heslop-Harrison, J. 1980. Darwin and the movement of plants: a retrospect. In *Plant growth substances 1979*, F. Skoog (ed.), 3–14. Berlin: Springer.

Hill, B. S. & G. P. Findlay 1981. The power of movement in plants: the role of osmotic machines. *Quarterly Review of Biophysics* **14**, 173–222.

Hilley, W. E. 1913. On the value of different degrees of centrifugal force as geotropic stimuli. *Annals of Botany* **27**, 719–58.

Hirouchi, T. & S. Suda 1975. Thigmotropism in the growth of pollen tubes of *Lilium longiflorum*. *Plant and Cell Physiology* **16**, 377–81.

Hodick, D. & A. Sievers 1988. The action potential of *Dionaea muscipula*. *Planta* **175**, 8–18.

Hofman, E. & E. Schäfer 1987. Red light-induced shift of the fluence response curve for first positive curvature of maize coleoptiles. *Plant and Cell Physiology* **28**, 37–45.

Hooker, H. D. 1915. Hydrotropism in roots of *Lupinus albus*. *Annals of Botany* **29**, 265–83.

Hosokawa, Y. & K. Kiyosawa 1983. Diurnal geotropic responses of the primary leaves of *Phaseolus vulgaris*. *Plant and Cell Physiology* **24**, 947–51.

Iino, M. 1987. Kinetic modelling of phototropism in maize coleoptiles. *Planta* **171**, 110–26.

Iino, M. & W. R. Briggs 1984. Growth distribution during first positive phototropic curvature of maize coleoptiles. *Plant, Cell & Environment* **7**, 97–104.

Iino, M., W. R. Briggs & E. Schäfer 1984. Phytochrome-mediated phototropism in maize seedling shoots. *Planta* **160**, 41–51.

Iino, M., E. Schäfer & W. R. Briggs 1985. Photoperception sites for phytochrome-mediated phototropism of maize coleoptiles. *Planta* **162**, 477–9.

Imaseki, H. 1985. Hormonal control of wound-induced responses. In *Hormonal regulation of development. III Role of environmental factors*, Encyclopaedia of Plant Physiology, NS Vol. 11, R. P. Pharis & D. M. Reid (eds), 484–512. Berlin: Springer.

Ishizawa, K. & S. Wada 1979. Growth and phototropic bending in *Boergesenia* rhizoid. *Plant and Cell Physiology* **20**, 973–87.

Iwami, S. & Y. Masuda 1976. Distribution of labelled auxin in geotropically stimulated stems of cucumber and pea. *Plant and Cell Physiology* **17**, 227–37.

Iwanami, Y. 1953. Physiological researches on pollen. *Botanical Magazine, Tokyo* **66**, 189–96.

Jackson, D. L. & J. A. McWha 1984. Grain or coleoptile tip as the source of IAA in cereal shoots? *Plant, Cell & Environment* **7**, 15–21.

Jackson, W. B. & P. W. Barlow 1981. Root geotropism and the role of growth regulators from the cap: a re-examination. *Plant, Cell & Environment* **4**, 107–23.

Jacobs, W. P. 1947. The development of the gynophore of the peanut plant *Arachis hypogea*. I The distribution of mitosis, the region of greatest elongation and the maintenance of vascular continuity in the intercalary meristem. *American Journal of Botany* **34**, 361–8.

Jaffe, L. & H. Etzold 1965. Tropic responses of *Funaria* spores to red light. *Biophysical Journal* **5**, 715–42.

Jaffe, M. J. 1973. Thigmomorphogenesis. The response of plant growth to mechanical stimulation. *Planta* **114**, 143–54.

Jaffe, M. J. 1975. The role of auxin in the early events of the contact coiling of tendrils. *Plant Science Letters* **5**, 217–25.

Jaffe, M. J. 1980. On the mechanism of contact coiling of tendrils. In *Plant growth substances 1979*, F. Skoog (ed.), 481–95. Berlin: Springer.

Jaffe, M. J. 1985. Wind and other mechanical effects in the development and behaviour of plants, with special emphasis on the role of hormones. In *Hormonal regulation of development. III Role of environmental factors*, Encyclopaedia of Plant Physiology, NS Vol. 11, R. P. Pharis & D. M. Reid (eds), 444–83. Berlin: Springer.

Jaffe, M. J. & A. W. Galston 1968. The physiology of tendrils. *Annual Review of Plant Physiology* **19**, 417–34.

Jaffe, M. J., C. Gibson & R. Biro 1977. Physiological studies of mechanically stimulated motor responses of flower parts. I Characterisation of the thigmotropic stamens of *Portulaca grandiflora*. *Botanical Gazette* **138**, 438–47.

Jaffe, M. J. et al. 1985. Computer-assisted image analysis of plant growth, thigmomorphogenesis and gravitropism. *Plant Physiology* **77**, 722–30.

Jankiewicz, L. S. 1971. Gravimorphism in higher plants. In *Gravity and the organism*, S. A. Gordon & M. J. Cohen (eds), 317–31. Chicago: University of Chicago Press.

Johnsson, A. 1979. Circumnutation. In *Physiology of movements*, Encyclopaedia of Plant Physiology, NS Vol. 7, W. Haupt & M. E. Feinleib (eds), 627–46. Berlin: Springer.

Johnsson, A. & D. G. Heathcote 1973. Experimental evidence and models on circumnutations. *Zeitschrift für Pflanzenphysiologie* **70**, 371–405.

Juniper, B. E. 1976. Geotropism. *Annual Review of Plant Physiology* **27**, 385–406.

Junker, S. 1976. Auxin transport in tendril segments of *Passiflora caerula*. *Physiologia Plantarum* **37**, 258–62.

Junker, S. 1977a. Ultrastructure and tactile papillae on tendrils of *Eccremocarpus scaber*. *New Phytologist* **78**, 607–10.

Junker, S. 1977b. Thigmonastic coiling of tendrils of *Passiflora quadrangularis* is not caused by lateral redistribution of auxin. *Physiologia Plantarum* **41**, 51–4.

Junker, S. & L. Reinhold 1975. A scanning electron microscope survey of the surface of sensitive tendrils. *Journal de Microscopie et de Biologie Cellulaire* **23**, 175–80.

Kadota, A., M. Wada & M. Furaya 1982. Phytochrome mediated phytotropism and different dichroic orientation of P_r and P_{fr} in protonemata of the fern *Adiantum capillus-veneris*. Photochemistry and Photobiology **35**, 533–6.

Kaldeway, H. 1962. Diageotropismus der spross und blätter einschliesslich epinastie, hyponastie, entfaltungsbewegungen. In *Physiology of movements*, Encyclopaedia of Plant Physiology, Vol. 17, part 2, W. Ruhland (ed.), 200–321. Berlin: Springer.

Kaldeway, H. 1971. Geoepinasty, an example of gravimorphism. In *Gravity and the organism*, S. A. Gordon & M. J. Cohen (eds), 333–9. Chicago: University of Chicago Press.

Kang, B. G. 1979. Epinasty. In *Physiology of movements*, Encyclopaedia of Plant Physiology, NS Vol. 7, W. Haupt & M. E. Feinleib (eds), 647–68. Berlin: Springer.

Kataoka, H. 1975. Phototropism in *Vaucheria geminata*. I The action spectrum. *Plant and Cell Physiology* **16**, 427–37.

Kataoka, H. & M. H. Weisenseel 1988. Blue light promotes ionic current influx at the growing apex of *Vaucheria terrestris*. *Planta* **173**, 490–9.

Kaufman, P. B. *et al.* 1987. How cereal grass shoots perceive and respond to gravity. *American Journal of Botany* **74**, 1446–57.

Knox, R. B. 1984. Pollen–pistil interactions. In *Cellular interactions*, Encyclopaedia of Plant Physiology, NS Vol. 17, H. F. Linskins & J. Heslop-Harrison (eds), 508–608. Berlin: Springer.

Kochert, G. 1978. Sexual pheromones in algae and fungi. *Annual Review of Plant Physiology* **29**, 461–86.

Kohji, J., K. Nishtani & Y. Masuda 1981. A study on the mechanism of nodding initiation of the flower stalk in a poppy. *Plant and Cell Physiology* **22**, 413–22.

Koller, D. 1986. The control of leaf orientation by light. *Photochemistry and Photobiology* **44**, 819–26.

Koller, D., I. Leviton & W. R. Briggs 1985. The vectorial photo-excitation in solar-tracking leaves of *Lavatera cretica*. *Photochemistry and Photobiology* **42**, 717–23.

Konjevic, R., D. Grubisic & M. Neskovic 1987. The effect of Norflurazon on the phototropic reaction in *Phaseolus vulgaris* seedlings. *Photochemistry and Photobiology* **45**, 821–4.

Koukkari, W. L. & S. B. Warde 1985. Rhythms and their relations to hormones. In *Hormonal regulation of development. III Role of environmental factors*, Encyclopaedia of Plant Physiology, NS Vol. 11, R. P. Pharis & D. M. Reid (eds), 37–78. Berlin: Springer.

Kutschera, L. 1960. *Wurzelatlas mitteleuropäischer ackerunkranter und kulturpflanzen*. Frankfurt-am-Main: DLG-Verlags-GmbH.

Kutschera, U. & W. R. Briggs 1987. Differential effect of auxin on *in vivo* extensibility of cortical cylinder and epidermis in pea internodes. *Plant Physiology* **84**, 1361–6.

Kutschera, U. & W. R. Briggs 1988. Growth, *in vivo* extensibility and tissue tension in developing pea internodes. *Plant Physiology* **86**, 306–11.

Kutschera, U., R. Bergfield & P. Schopfer 1987. Cooperation of epidermis and inner tissues in auxin-mediated growth of maize coleoptiles. *Planta* **170**, 168–80.

Lachno, D. R., R. S. Harrison-Murray & L. J. Audus 1982. The effects of mechanical impedance to growth on the levels of ABA and IAA in root tips of *Zea mays*. *Journal of Experimental Botany* **33**, 943–51.

Langham, D. G. 1941. The effect of light on the growth habit of plants. *American Journal of Botany* **20**, 951–6.

Larsen, P. 1962a. Geotropism. An introduction. In *Physiology of movements*, Encyclopaedia of Plant Physiology, Vol. 17, part 2, W. Ruhland (ed.), 34–73. Berlin: Springer.

Larsen, P. 1962b. Orthogeotropism in roots. In *Physiology of movements*, Encyclopaedia of Plant Physiology, Vol. 17, part 2, W. Ruhland (ed.), 153–99. Berlin: Springer.

Lee, J. S., T. J. Mulkey & M. L. Evans 1983. Reversible loss of gravitropic sensitivity in maize roots after tip application of calcium chelators. *Science* **220**, 1375–6.

Lee, J. S., T. J. Mulkey & M. L. Evans 1984. Inhibition of polar calcium movement and gravitropism in roots treated with auxin transport inhibitors. *Planta* **160**, 536–43.

Leshen, Y. Y. 1977. Sunflower: a misnomer? *Nature* **269**, 102.

Lichtner, F. T. & S. E. Williams 1977. Prey capture and factors controlling trap narrowing in *Dionaea* (Droseraceae). *American Journal of Botany* **64**, 881–6.

Lintilhac, P. M. & T. B. Vesecky 1981. Mechanical stress and cell wall orientation in plants. II The application of controlled directional stress to growing plants; with a discussion on the nature of the wound reaction. *American Journal of Botany* **68**, 1222–30.

Lintilhac, P. M. & T. B. Vesecky 1984. Stress-induced alignment of division plane in plant tissues grown *in vitro*. *Nature* **307**, 363–4.

Lloyd, F. E. 1942. *The carnivorous plants*. New York: Ronald.

Loomis, W. E. & L. M. Ewan 1936. Hydrotropic responses of roots in soil. *Botanical Gazette* **97**, 728–43.

Löser, G. & E. Schäfer 1980. Phototropism in *Phycomyces*: a photochromic pigment? In *The blue light syndrome*, H. Senger (ed.), 244–50. New York: Springer.

Löser, G. & E. Schäfer 1986. Are there several photoreceptors involved in phototropism of *Phycomyces blakesleeanus*? Kinetic studies of dichromatic irradiation. *Photochemistry and Photobiology* **43**, 195–204.

MacDonald, I. R., J. W. Hart 1985a. The role of the apex in normal and tropic growth in sunflower hypocotyls. *Planta* **163**, 549–53.

MacDonald, I. R. & J. W. Hart 1985b. Apical involvement in geogrowth response of the horizontal hypocotyl. *Journal of Plant Physiology* **118**, 353–6.

MacDonald, I. R. & J. W. Hart 1987a. New light on the Cholodny–Went theory. *Plant Physiology* **84**, 568–70.

MacDonald, I. R. & J. W. Hart 1987b. Tropisms as indicators of hormone mediated growth phenomena. In *Hormones in plant development: a critical appraisal*, N. V. Hoad, J. R. Lenton, R. Atkin & M. B. Jackson (eds), 231–49. London: Butterworth.

MacDonald, I. R., D. C. Gordon & J. W. Hart 1987. Cyclamen coiling: the migration of a growth response. *Plant, Cell & Environment* **10**, 613–18.

MacDonald, I. R., J. W. Hart & D. C. Gordon 1983b. Analysis of growth during geotropic curvature in seedling hypocotyls. *Plant, Cell & Environment* **6**, 401–6.

MacDonald, I. R., D. C. Gordon, J. W. Hart & E. P. Maher 1983a. The positive hook: the role of gravity in the formation and opening of the apical hook. *Planta* **158**, 76–81.

McIntyre, G. I. 1980. The role of water distribution in plant tropisms. *Australian Journal of Plant Physiology* **7**, 401–13.

MacLeod, K., J. Digby & R. D. Firn 1985. Evidence inconsistent with the Blaauw model of phototropism. *Journal of Experimental Botany* **36**, 312–19.

MacLeod, K., R. D. Firn & J. Digby 1986. The phototropic responses of *Avena* coleoptiles. *Journal of Experimental Botany* **37**, 542–8.

McNab, R. M. 1979. Chemotaxis in bacteria. In *Physiology of movements*, Encyclopaedia of Plant Physiology, Vol. 7, W. Haupt & M. E. Feinleib (eds), 310–34. Berlin: Springer.

Mandoli, D. F. & W. R. Briggs 1982. The photoreceptive sites and the function of tissue light-piping in photomorphogenesis of etiolated oat seedlings. *Plant, Cell & Environment* **5**, 137–45.

Mandoli, D. F., J. Tepperman, E. Huala & W. R. Briggs 1984. Photobiology of diagravitropic maize roots. *Plant Physiology* **75**, 359–63.

Mayer, W., I. Moser & E. Bünning 1973. The epidermis as the site of light perception in circadian leaf movement and photoperiod induction. *Zeitschrift für Pflanzenphysiologie* **70**, 66–73.

Mayer, W.-E., D. Flach, M. V. S. Rajn & E. Wiech 1985. Mechanics of circadian pulvini movements in *Phaseolus coccineus*. *Planta* **163**, 381–90.

Mertens, R. & E. W. Weiler 1983. Kinetic studies on the redistribution of endogenous growth regulators in gravireacting plant organs. *Planta* **158**, 339–48.

Meudt, W. J. 1987. Investigations on the mechanism of the brassinosteroid response. VI Effect of brassinolide on gravitropism of bean hypocotyls. *Plant Physiology* **83**, 195–8.

Migliaccio, F. & A. W. Galston 1987. On the nature and origin of the calcium asymmetry arising during gravitropic response in etiolated pea epicotyls. *Plant Physiology* **85**, 542–8.

Miller, A. L., E. Shand & N. A. A. Gow 1988. Ion currents associated with root tips, emerging laterals and induced wound sites in *Nicotiana tabacum*: spatial relationship proposed between electrical fields and phytophthoran zoospore infection. *Plant, Cell & Environment* **11**, 21–5.

Miller, A. L., J. A. Raven, J. I. Sprent & M. H. Weisenseel 1986. Endogenous ion currents traverse growing roots and root hairs of *Trifolium repens*. *Plant, Cell & Environment* **9**, 79–83.

Miller, J. H. 1968. Fern gametophytes as experimental material. *Botanical Review* **34**, 361–440.

Millet, B., D. Melin & P. Badot 1987. Circumnutation: a model for signal transduction from cell to cell. In *The cell surface in signal transduction*, E. Wagner, H. Greppin & B. Millet (eds), 169–79. Berlin: Plenum Press.

Mitchell, J. E. 1976. The effect of roots on the activity of soil-borne root pathogens. In *Physiological plant pathology*, Encyclopaedia of Plant Physiology, NS Vol. 4, R. Heitefuss & P. H. Williams (eds), 94–128. Berlin: Springer.

Mohr, H. 1972. *Lectures in photomorphogenesis*. Berlin: Springer.

Montaldi, E. R. 1969. Gibberellin-sugar interaction regulating the growth habit of Bermuda grass. *Experientia* **25**, 91–2.

Montaldi, E. R. 1979. Thigmogeotropism as a cause of rhizome absence in *Cynodon plectostachyum*, and depth distribution in *C. dactylon*. *Phyton (Buenos Aires)* **37**, 21–4.

Moore, R. & M. L. Evans 1986. How roots perceive and respond to gravity. *American Journal of Botany* **73**, 574–87.

Moore, R. & J. Pasieniuk 1984. Graviresponsiveness and the development of columella tissue in primary and lateral roots of *Ricinus communis*. *Plant Physiology* **74**, 529–33.

Mueller, W. J., F. B. Salisbury & P. T. Blotter 1984. Gravitropism in higher plant shoots. II Dimensional and pressure changes during stem bending. *Plant Physiology* **76**, 993–9.

Naitoh, Y. 1984. Mechanosensory transduction in protozoa. In *Membranes and sensory transduction*, G. Colombetti & F. Lenci (eds), 113–36. New York: Plenum Press.

Nebel, B. J. 1968. Action spectra for photogrowth and phototropism in protonemata of the moss *Physcomitrium turbinatum*. *Planta* **81**, 287–302.

Neel, P. L. & R. W. Harris 1971. Motion-induced inhibition of elongation and induction of dormancy in *Liquidamber*. *Science* **173**, 58–9.

Neuscheler-Wirth, H. 1970. Photomorphogenese und phototropismus bei *Mougeotia*. *Zeitschrift für Pflanzenphysiologie* **63**, 238–60.

Newcombe, F. C. & A. L. Rhodes 1904. Chemotropism of roots. *Botanical Gazette* **37**, 23–35.

Ney, D. & P.-E. Pilet 1981. Nutation of growing and georeacting roots. *Plant, Cell & Environment* **4**, 339–43.

Nick, P. & E. Schäfer 1988. Interaction of gravi- and phototropic stimulation in the response of maize (*Zea mays*) coleoptiles. *Planta* **173**, 213–20.

Njus, D., F. M. Sulzman & J. W. Hastings 1974. Membrane model for the circadian clock. *Nature* **248**, 116–20.

Öquist, G. 1983. Effects of low temperature on photosynthesis. *Plant, Cell & Environment* **6**, 281–300.

Otto, M. K., M. Jayaram, R. M. Hamilton & M. Delbrück 1981. Replacement of riboflavin by an analogue in the blue-light photoreceptor of *Phycomyces*. *Proceedings of the National Academy of Sciences (USA)* **78**, 266–9.

Overbeek, J. van 1933. Wuchsstaff, lichtwachstumreaktion und phototropismus bei *Raphanus*. *Recueil des Travaux des Société Botanica Neerlandica* **30**, 537–626.

Page, R. M. 1968. Phototropism in fungi. In *Photophysiology*, Vol. 3, A. C. Giese (ed.), 65–90. New York: Academic Press.

Palmer, J. H. 1956. The nature of the growth response to sunlight shown by certain stoloniferous and prostrate tropical grasses. *New Phytologist* **55**, 346–55.

Palmer, J. H. 1958. Studies in the behaviour of the rhizome of *Agropyron repens*. I The seasonal development and growth of the parent plant and rhizome. *New Phytologist* **57**, 145–52.

Palmer, J. H. 1985. Epinasty, hyponasty and related topics. In *Hormonal regulation of development. III Role of environmental factors*, Encyclopaedia of Plant Physiology, NS Vol. 11, R. P. Pharis & D. M. Reid (eds), 139–68. Berlin: Springer.

Pappas, T. & C. A. Mitchell 1985. Effects of seismic stress on the vegetative growth of *Glycine max*. *Plant, Cell & Environment* **8**, 143–8.

Parker, M. L. 1979. Morphology and ultrastructure of the gravity-sensitive leaf sheath base of the grass *Echinochloa colonum*. *Planta* **145**, 471–7.

Parks, B. M. & K. L. Poff 1985. Phytochrome conversion as an *in situ* assay for effective light gradients in etiolated seedlings of *Zea mays*. *Photochemistry and Photobiology* **41**, 317–22.

Parsons, A., K. Macleod, R. D. Firn & J. Digby 1984. Light gradients in shoots subjected to unilateral illumination: implications for phototropism. *Plant, Cell & Environment* **7**, 325–32.

Parsons, A., R. D. Firn & J. Digby 1988. The role of the coleoptile apex in controlling organ elongation. I The effects of decapitation and apical incisions. *Journal of Experimental Botany* **39**, 1331–41.

Paterson, N. W., J. D. B. Weyers & H. Schildknecht 1987. The effects of a turgorin on stomatal movement and transpiration in *Commelina communis*. *Journal of Plant Physiology* **128**, 491–5.

Pfeffer, W. 1906. *The physiology of plants. III Movement*, translated by A. J. Ewart. Oxford: Clarendon Press.

Phillips, I. D. J. 1972a. Endogenous gibberellin transport and biosynthesis in relation to geotropic induction in excised sunflower shoot tips. *Planta* **105**, 234–44.

Phillips, I. D. J. 1972b. Diffusible gibberellins and phototropism in *Helianthus annuus*. *Planta* **106**, 363–7.

Pickard, B. G. 1969. Second positive phototropic response patterns of the oat coleoptile. *Planta* **88**, 1–33.

Pickard, B. G. 1973. Action potentials in higher plants. *Botanical Review* **39** 172–201.

Pickard, B. G. 1980. Comment in the discussion of the paper by Gradmann & Mummerts (1980) Plant action potentials. In *Plant membrane transport: current conceptual issues*, R. M. Spanswick, W. J. Lucas & J. Dainty (eds), 233–47. Amsterdam: Elsevier.

Pickard, B. G. 1985a. Roles of hormones, protons and calcium in geotropism. In *Hormonal regulation of development. III Role of environmental factors*, Encyclopaedia of Plant Physiology, NS Vol. 11, R. P. Pharis & D. M. Reid (eds), 193–281. Berlin: Springer.

Pickard, B. G. 1985b. Roles of hormones in phototropism. In *Hormonal regulation of development. III Role of environmental factors*, Encyclopaedia of Plant Physiology, NS Vol. 11, R. P. Pharis & D. M. Reid (eds), 364–417. Berlin: Springer.

Pickard, B. G. 1985c. Early events in geotropism of seedling shoots. *Annual Review of Plant Physiology* **36**, 55–75.

Pickard, B. G. & K. V. Thimann 1966. Geotropic response of wheat coleoptiles in absence of amyloplast starch. *Journal of General Physiology* **49**, 1065–86.

Pilet, P.-E. & L. Rivier 1981. Abscisic acid distribution in horizontal maize root segments. *Planta* **153**, 453–8.

Poff, K. L. 1983. Perception of a unilateral light stimulus. *Philosophical Transactions of the Royal Society of London* **B303**, 479–87.

Poff, K. L., D. R. Fontana & B. D. Whitaker 1984. Temperature sensing in micro-organisms. In *Membranes and sensory transduction*, G. Colombetti & F. Lenci (eds), 137–62. New York: Plenum Press.

Pohl, U. & V. E. A. Russo 1984. Phototropism. In *Membranes and sensory transduction*, G. Colombetti & F. Lenci (eds), 231–329. New York: Plenum Press.

Pollard, E. C. 1971. Physical determinants of receptor mechanism. In *Gravity and organism*, S. A. Gordon & M. J. Cohen (eds), 25–34. Chicago: University of Chicago Press.

Pope, D. G. 1982. Effect of peeling on IAA-induced growth in *Avena* coleoptiles. *Annals of Botany* **49**, 493–501.

Presti, D. E. 1983. The biology of carotenes and flavins. In *The biology of photoreception*, SEB Symposium No. 36, D. Cosens & D. Vince-Prue (eds), 133–80. Cambridge: Cambridge University Press.

Prichard, J. M. & I. N. Forseth 1988. Photosynthetic responses of two heliotropic legumes from contrasting sites. *Plant, Cell & Environment* **11**, 591–601.

Ransom, J. S. & R. Moore 1983. Geoperception in primary and lateral roots of *Phaseolus vulgaris*. I Structure of columella cells. *American Journal of Botany* **70**, 1048–56.

Reches, S., Y. Leshem & J. Wurtzburger 1974. On hormones and weeping: asymmetric hormone distribution and the pendulous growth habit of the weeping mulberry. *New Phytologist* **73**, 841–6.

Reinhold, L., T. Sachs & L. Vislovska 1970. The role of auxin in thigmotropism. In *Plant growth substances*, D. J. Carr (ed.), 731–7. Berlin: Springer.

Reynolds-Green, J. 1909. *A history of botany (1860–1900)*. Oxford: Clarendon Press.

Rich, T. C. G., G. C. Whitelam & H. Smith 1985. Phototropism and axis extension in light-grown mustard (*Sinapis alba*) seedlings. *Photochemistry and Photobiology* **42**, 789–92.

Rich, T. C. G., G. C. Whitelam & H. Smith 1987. Analysis of growth rates during phototropism: modifications by separate light-growth responses. *Plant, Cell & Environment* **10**, 303–11.

Roblin, G. 1979. *Mimosa pudica*: a model for the study of excitability in plants. *Biological Reviews* **54**, 135–53.

Roblin, G. & P. Fleurat-Lessard 1987. Redistribution of potassium, chloride and calcium during the gravitropically induced movement of *Mimosa pudica* pulvinus. *Planta* **170**, 242–8.

Rosen, W. G. 1961. Studies on pollen tube chemotropism. *American Journal of Botany* **48**, 889–95.

Rosen, W. G. 1971. Pistil–pollen interactions in *Lilium*. In *Pollen development and physiology*, J. Heslop-Harrison (ed.), 239–54. London: Butterworth.

Rufelt, H. 1962. Plagiogeotropism in roots. In *Physiology of movements*, Encyclopaedia of Plant Physiology, Vol. 17, part 2, W. Ruhland (ed.), 322–43. Berlin: Springer.

Rufelt, H. 1969. Geo- and hydrotropic responses of roots. In *Root growth*, W. J. Whittington (ed.), 54–64. London: Butterworth.

Russell, R. S. 1977. *Plant root systems*. Maidenhead, England: McGraw-Hill.

Russo, V. A. E. 1980. Sensory transduction in phototropism: genetic and physiological analysis in *Phycomyces*. In *Photoreception and sensory transduction in aneural organisms*, F. Lenci & G. Colombetti (eds), 373–95. New York: Plenum Press.

Ryan, C. A. 1987. Oligosaccharide signalling in plants. *Annual Review of Cell Biology* **3**, 295–317.

Sachs, J. 1887. *Lectures on the physiology of plants*, translated by H. M. Ward. Oxford: Clarendon Press.

Sachs, J. 1890. *A history of botany (1530–1860)*. Oxford: Clarendon Press.

Sack, F. D., M. M. Suyemoto & A. C. Leopold 1986. Amyloplast sedimentation and organelle saltation in living corn columella cells. *American Journal of Botany* **73**, 1692–8.

Salisbury, F. B. & C. W. Ross 1985. *Plant Physiology*, 3rd edn. Belmont, California: Wadsworth.

Salisbury, F. B., P. Rorabaugh, R. White & L. Gillespie 1986. A key role for sensitivity to auxin in gravitropic stem bending. *Plant Physiology* **80**, S–26.

Satter, R. L. 1979. Leaf movements and tendril coiling. In *Physiology of movements*, Encyclopaedia of Plant Physiology, NS Vol. 7, W. Haupt & M. E. Feinleib (eds), 442–84. Berlin: Springer.

Schildknecht, K. 1984. Turgorins: new chemical messengers for plant behaviour. *Endeavour* **8**, 113–6.

Schmidt, W. 1983. The physiology of blue light systems. In *The biology of photoreception*, SEB Symposium No. 36, D. Cosens & D. Vince-Prue (eds), 305–30. Cambridge: Cambridge University Press.

Schrank, A. R. 1959. Electronasty and electrotropism. In *Physiology of movements*, Encyclopaedia of Plant Physiology, Vol. 17, part 2, W. Ruhland (ed.), 148–63. Berlin: Springer.

Schulze, A. & R. S. Bandurski 1986. Movement of IAA from stele to cortex of *Zea mays* seedlings. *Plant Physiology* **80**, S–26.

Schulze, A. & R. S. Bandurski 1987. A gravity-induced asymmetric unloading of IAA from the stele of *Zea mays* into the mesocotyl cortex. *Plant Physiology* **83**, S–102.

Schwabe, W. W. 1968. Studies on the role of the leaf epiderm in photoperiodic perception in *Kalanchoe blossfeldiana*. *Journal of Experimental Botany* **19**, 108–13.

Selker, J. M. L. & A. Sievers 1987. Analysis of extension and curvature during the graviresponse in *Lepidium* roots. *American Journal of Botany* **74**, 1863–71.

Senger, H. & W. R. Briggs 1981. The blue light receptor(s): primary actions and subsequent metabolic changes. In *Photochemical and Photobiological Review*, vol. 6, K. C. Smith (ed.), 1–38. New York: Plenum Press.

Shen-Miller, J. & S. A. Gordon 1967. Gravitational compensation and the phototropic response of oat coleoptiles. *Plant Physiology* **42**, 352–60.

Shen-Miller, J., R. Hinchman & S. A. Gordon 1968. Thresholds for georesponse to acceleration in gravity-compensated *Avena* seedlings. *Plant Physiology* **43**, 338–44.

Sherriff, D. W. & M. M. Ludlow 1985. Diaheliotropic responses of leaves of *Macroptilium atropurpureum*. *Australian Journal of Plant Physiology* **12**, 151–71.

Shropshire, W. 1962. The lens effect and phototropism of *Phycomyces*. *Journal of General Physiology* **45**, 949–58.

Shropshire, W. 1979. Stimulus perception. In *Physiology of movements*, Encyclopaedia of Plant Physiology, NS Vol. 7, W. Haupt & W. E. Feinleib (eds), 10–41. Berlin: Springer.

Shropshire, W. & H. Mohr 1970. Gradient formation of anthocyanin in seedlings of *Fagopyrum* and *Sinapis* unilaterally exposed to red and far-red light. *Photochemistry and Photobiology* **12**, 143–9.

Shuttleworth, J. E. & M. Black 1977. The role of the cotyledons in phototropism of de-etiolated seedlings. *Planta* **135**, 51–5.

Sievers, A. & D. Volkmann 1979. Gravitropism in single cells. In *Physiology of movements*, Encyclopaedia of Plant Physiology, NS Vol. 7, W. Haupt & M. E. Feinleib (eds), 567–72. Berlin: Springer.

Sievers, A., H. M. Berhens, T. J. Buckout & D. Gradman 1984. Can a calcium pump in the endoplasmic reticulum of the *Lepidium* root be the trigger for rapid changes in membrane potential after gravistimulation? *Zeitschrift für Pflanzenphysiologie* **114**, 195–200.

Silk, W. K. 1984. Quantitative descriptions of plant development. *Annual Review of Plant Physiology* **35**, 479–518.

Simons, P. J. 1981. The role of electricity in plant movements. *New Phytologist* **87**, 11–37.

Sliwinski, J. E. & F. B. Salisbury 1984. Gravitropism in higher plant shoots. III Cell dimensions during gravitropic bending: perception of gravity. *Plant Physiology* **76**, 1000–8.

Slocum, R. D. & R. D. Roux 1983. Cellular and sub-cellular localisation of calcium in gravistimulated oat coleoptiles and its possible significance in the establishment of tropic curvature. *Planta* **157**, 481–92.

Smith, H. 1982. Light quality, photoperception and plant strategy. *Annual Review of Plant Physiology* **33**, 481–518.

Smith, H. 1984. Plants that track the sun. *Nature* **308**, 774.

Snow, R. 1962. Geostrophism. In *Physiology of movements*, Encyclopaedia of Plant Physiology, Vol. 17, part 2, W. Ruhland (ed.), 378–89. Berlin: Springer.

Spalding, V. M. 1894. The traumatropic curvature of roots. *Annals of Botany* **8**, 423–51.

Steiner, A. M. 1969. Action spectrum for polarotropism of the germ tube of the liverwort *Sphaerocarpus donnellii*. *Planta* **86**, 343–52.

Steinitz, B. & K. L. Poff 1986. A single positive phototropic response induced with pulsed light in hypocotyls of *Arabidopsis thaliana* seedlings. *Planta* **168**, 305–15.

Steinitz, B., Z. Ren & K. L. Poff 1985. Blue and green light-induced phototropism in *Arabidopsis* and *Lactuca* seedlings. *Plant Physiology* **77**, 248–51.

Suzuki, T., N. Kando & T. Fujii 1979. Distribution of growth regulators in relation to the light-induced geotropic responsiveness in *Zea* roots. *Planta* **145**, 323–9.

Tanada, T. & C. Vinten-Johansen 1980. Gravity induces fast electrical field change in soybean hypocotyls. *Plant, Cell & Environment* **3**, 127–30.

Taylor, A., J. W. Hart & A. R. Berg 1988. Different forms of phototropic response induced by microbeam irradiation of discrete regions of an organ. *Plant, Cell & Environment* **11**, 645–52.

Taylor, B. L. & S. M. Panasenko 1984. Biochemistry of chemosensory behaviour in prokaryotes and unicellular eukaryotes. In *Membranes and sensory transduction*, G. Colombetti & F. Lenci (eds), 71–112. New York: Plenum Press.

Thimann, K. V. & C. L. Schneider 1938. Differential growth in plant tissues. *American Journal of Botany* **25**, 627–41.

Trewavas, A. J. 1981. How do plant growth substances work? *Plant, Cell & Environment* **4**, 203–28.

Trewavas, A. J. 1982. Growth substance sensitivity: the limiting factor in plant development. *Physiologica Plantarum* **55**, 60–72.

Trewavas, A. J. 1986. Understanding the control of plant development and the role of hormones. *Australian Journal of Plant Physiology* **13**, 447–57.

Tsuboi, C. 1983. *Gravity*. London: Allen & Unwin.

Ullrich, C.-H. 1978. Continuous measurement of initial curvature of maize coleoptiles induced by lateral auxin application. *Planta* **140**, 201–11.

Vierstra, R. D. & K. L. Poff 1981a. Role of carotenoids in the phototropic response of corn seedlings. *Plant Physiology* **68**, 798–801.

Vierstra, R. D. & K. L. Poff 1981b. Mechanism of specific inhibition of photo-tropism by phenylacetic acid in corn seedlings. *Plant Physiology* **67**, 1011–15.

Vogelmann, T. C. 1984. Site of light perception and motor cells in a sun-tracking lupin. (*Lupinus succulentus*). *Physiologica Plantarum* **62**, 335–40.

Vogelmann, T. C. & W. Haupt 1985. The blue light gradient in unilaterally irradiated maize coleoptiles: measurement with a fibre optic probe. *Photochemistry and Photobiology* **41**, 569–76.

Volkmann, D. & A. Sievers 1979. Graviperception in multicellular organs. In *Physiology of movements*, Encyclopaedia of Plant Physiology, NS Vol. 7, W. Haupt & M. E. Feinleib (eds), 573–600. Berlin: Springer.

Weisenseel, M. H. 1979. Induction of polarity. In *Physiology of movements*, Encyclopaedia of Plant Physiology, NS Vol. 7, W. Haupt & M. E. Feinleib (eds), 485–505. Berlin: Springer.

Went, F. A. F. C. 1935. The investigations on growth and tropisms carried on in the botanical laboratory of the University of Utrecht during the last decade. *Biological Reviews* **10**, 187–207.

Went, F. W. 1974. Reflections and speculations. *Annual Review of Plant Physiology* **25**, 1–26.

Went, F. W. & K. V. Thimann 1937. *Phytohormones*. New York: MacMillan.

Werk, K. S. & J. Ehleringer 1984. Non-random leaf orientation in *Lactuca serriola*. *Plant, Cell & Environment* **7**, 81–7.

Westing, A. H. 1971. A case against statoliths. In *Gravity and the organism*, S. A. Gordon & M. J. Cohen (eds), 97–102. Chicago: University of Chicago Press.

Weyers, J. D. B., N. W. Paterson & R. A'Brook 1987. Towards a quantitative definition of plant hormone sensitivity. *Plant, Cell & Environment* **10**, 1–10.

Wheeler, R. M. & F. B. Salisbury 1981. Gravitropism in higher plant shoots. I A role for ethylene. *Plant Physiology* **67**, 686–90.

Whittaker, R. 1975. *Communities and ecosystems*, 2nd edn. New York: MacMillan.

Wiese, L. 1984. Mating systems in unicellular algae. In *Cellular interactions*, Encyclopaedia of Plant Physiology, NS Vol. 17, H. F. Linskins & J. Heslop-Harrison (eds), 238–60. Berlin: Springer.

Wilkins, M. B. 1979. Growth control mechanisms in gravitropism. In *Physiology of movements*, Encyclopaedia of Plant Physiology, NS Vol. 7, W. Haupt & M. E. Feinleib (eds), 601–26. Berlin: Springer.

Wilkins, M. B. 1984. Gravitropism. In *Advanced plant physiology*, M. B. Wilkins (ed.), 163–85. London: Pitman.

Wilson, B. F. 1973. White pine shoots: roles of gravity and epinasty in movements and compression wood location. *American Journal of Botany* 60, 597–601.

Wilson, B. F. & R. R. Archer 1977. Reaction wood: induction and mechanical action. *Annual Review of Plant Physiology* 28, 24–43.

Witzum, A. & M. Gersani 1975. The role of polar movement of IAA in the development of the peg in *Cucumis sativus*. *Botanical Gazette* 136, 5–16.

Woitzik, F. & H. Mohr 1988a. Control of hypocotyl phototropism by phytochrome in a dicotyledonous seedling (*Sesamum indicum*). *Plant, Cell & Environment* 11, 653–61.

Woitzik, F. & H. Mohr 1988b. Control of hypocotyl gravitropism by phytochrome in a dicotyledonous seedling (*Sesamum indicum*). *Plant, Cell & Environment* 11, 663–8.

Wright, M. 1986. The acquisition of gravisensitivity during the development of nodes of *Avena fatua*. *Journal of Plant Growth Regulation* 5, 37–47.

Wright, M., D. M. A. Monsdale & D. J. Osborne 1978. Evidence for a gravity regulated level of endogenous auxin controlling cell elongation and ethylene production during geotropic bending in grass nodes. *Biochemie und Physiologie der Pflanzen* 172, 581–96.

Wynn, W. K. 1976. Appressorium formation over stomates by the bean rust fungus: response to a surface contact stimulus. *Phytopathology* 66, 708–14.

Yin, H. C. 1941. Studies on the nyctinastic movement of the leaves of *Carica papaya*. *American Journal of Botany* 28, 250–61.

Zeiger, E., M. Iino & T. Ogawa 1985. The blue light response of stomata: pulse kinetics and some mechanistic implications. *Photochemistry and Photobiology* 42, 759–63.

Ziegler, H. 1962a. Chemotropismus. In *Physiology of movements*, Encyclopaedia of Plant Physiology, Vol. 17, part 2, W. Ruhland (ed.), 396–431. Berlin: Springer.

Ziegler, H. 1962b. Hydrotropismus. In *Physiology of movements*, Encyclopaedia of Plant Physiology, Vol. 17, part 2, W. Ruhland (ed.), 432–50. Berlin: Springer.

Zimmerman, B. K. & W. R. Briggs 1963. A kinetic model for the phototropic response of oat coleoptiles. *Plant Physiology* 38, 253–61.

Subject Index

204

Species Index